T0134006

CLAY SEALS OF OIL AND GAS DEPOSITS

Clay Seals of Oil and Gas Deposits

V.I. OSIPOV
V.N. SOKOLOV
V.V. EREMEEV

A.A. BALKEMA PUBLISHERS / LISSE / ABINGDON / EXTON (PA) / TOKYO

Library of Congress Cataloging-in-Publication Data

Typesetting: Charon Tec Pvt. Ltd, Chennai, India
Printed in the Netherlands by Krips The Print Force, Meppel

Published by: A.A. Balkema Publishers, a member of Swets & Zeitlinger Publishers
www.balkema.nl and www.szp.swets.nl

ISBN 90 5809 583 5

Contents

Introduction   VII

PART I   FORMATION OF CLAY SEAL PROPERTIES: THEORETICAL   1
              FUNDAMENTALS

Chapter 1   Composition of Clay Sediments and Their Structure   3
              Formation in Sedimentogenesis
              1.1. Composition of clay sediments   3
              1.2. Structure of clay sediments   7
              1.3. The theory of structure formation in clay dispersions   17
              1.4. Structural bonds   26
              1.5. Formation of primary microstructures in clay sediments   30

Chapter 2   Lithogenesis of Clay Sediments   37
              2.1. General information   37
              2.2. The stages of lithogenesis   38
              2.3. Boundaries between the stages and substages of lithogenesis   64

Chapter 3   Formation of the Properties of Clay Seals in Lithogenesis   71
              3.1. Density and porosity   71
              3.2. Water content and consistence   83
              3.3. Permeability   90
              3.4. Physical-mechanical properties   97
              3.5. Rheological properties   109
              3.6. General regularities in the formation of clay properties   112
                    in the process of lithogenesis

PART II   CHARACTERISTIC OF THE FACIES TYPES OF CLAY SEALS   119

Chapter 4   Depositional Environments of Clay Seal Formation   121
              4.1. The cyclic structure of sedimentary basin sequences   121
              4.2. Source provinces   123
              4.3. Evolution of sedimentation basins   125
              4.4. Classification of clay seals according to their depositional   131
                    environments and screening properties

Chapter 5   Facies Analysis of Clay Seals in Oil- and Gas-Bearing Basins   137
              5.1. The principal criteria for facies analysis   137
              5.2. Facies analysis of clay seals in sedimentary basins   139

Chapter 6   The Composition of the Facies Types of Clay Seals and Its Change   147
with Depth
     6.1. The composition of the facies types of clay seals   147
     6.2. A comparative characteristic of the composition of clay seals   158
         of various facies types

Chapter 7   Structure and Properties of the Facies Types of Clay Seals   163
     7.1. The influence of depositional environments on the   163
         structure formation
     7.2. Porosity   172
     7.3. Microfracturing   181
     7.4. Permeability   188
     7.5. Destruction of clay seals of different facies types with depth   190

PART III   MODELING OF THE CHANGE IN THE CLAY SEAL   195
PROPERTIES IN LITHOGENESIS

Chapter 8   Physicochemical Models of Clay Rock Microstructures   197
     8.1. The concepts of the physicochemical mechanics of clay rocks   197
         as a basis for modeling
     8.2. Calculation of the number of contacts   198
     8.3. Strength of individual contacts between clay particles   204

Chapter 9   Development of a Mathematical Model for Estimating the Strength   211
of Clay Seals
     9.1. Theoretical calculation of the change in the strength of   211
         individual contacts with depth
     9.2. Calculation of the change in the number of contacts in   225
         clay seals with depth
     9.3. Theoretical calculation of tensile and   229
         compressive strength of clay seals

Chapter 10   Comparison of Theoretical Calculation Results and Experimental Data   239
on the Mechanical Strength of Clay Seals
     10.1. Experimental data base   239
     10.2. Comparison of theoretical and experimental data   239

References   259

Subject Index   275

Introduction

Clays and clay rocks are the most common type of deposits in the Earth's sedimentary cover: they make up nearly half of the sediment mass in the lithosphere. Their volume in the Earth's crust amounts to 270 million km^3. Up to 85% of the total carbonaceous organic matter is associated with clays, including the argillaceous cement in clastic and carbonate rocks [1].

The evolution of clay deposits and organic matter in sedimentary basins in the course of geological history is closely interconnected. An analysis of sedimentogenesis in marine and oceanic basins has shown that clay material and organic matter are deposited together. Within shallow-water areas of intracontinental seas, oceanic shelf and continental slope, the fine grained clay oozes are enriched in organic matter of detrital origin. This is caused by the similar sizes of clay particles and organic matter and by specific nature of the former. In relatively deep-water clay sediments, the bulk of organic matter is of the sapropel type that has the greatest oil generation potential.

Clays and clay rocks play the most important role not only in the genesis of petroleum hydrocarbons but also in their concentration, as they form adequate screens for pools. Regional clay screens – fluid-tight sequences – are one of the leading factors controlling, under the conditions of wide-scale inter-reservoir migration of hydrocarbon fluids, the concentration of oil and gas pools within different intervals of the sedimentary cover in oil- and gas-bearing basins. Therefore, not only the development of reservoir rocks, but also the presence of clay horizons acting as screens is among the most important criteria in the search for oil and gas in sedimentary basins.

A connection of petroleum-productive sequences with the presence of overlying clay deposits of predominantly montmorillonitic composition has been established for the major oil- and gas-bearing basins of the Earth. This evidences for an important role of smectite components in the formation of regional clay screens-seals and shows that it is necessary to study the geological history of a basin and the depositional environments controlling the formation of clay sediments of different mineral and grain-size composition.

It is no less important to study the post-sedimentation alteration of clay deposits as they become buried and overlain by new horizons of sedimentary material. The composition and structure of clay deposits, their physical and mechanical properties are altered in the course of lithogenesis, with clays being gradually transformed into clay rocks and shales, and then into argillites and schists. These alterations are accompanied by the change in the screening properties of clay sequences, thus affecting the conditions of hydrocarbon localization and migration and being one of the critical factors controlling oil and gas reserves. Therefore, to predict the presence of hydrocarbon accumulations, one has to know the regularities in the alteration of the screening properties of clay seals with depth related to their porosity and microfracturing. Both of these parameters are determined, in their turn, by the degree of compaction and the strength properties of clay deposits.

The present work consists of 3 parts. The first part deals with theoretical principles of the formation of clay seal properties. It discusses the processes of structure formation and transformation of clay sediments in the course of lithogenesis and describes on this basis the laws governing the formation of their physical, filtration, and mechanical properties.

The second part describes conditions controlling the formation of clay sequences and gives the facies principle of classification for the formed clay seals. Based on the facies analysis of different oil- and gas-bearing basins, the facies types of seals are distinguished and correlation between the properties of clay deposits and the depth of their occurrence is analyzed.

The third, final part of the work deals with the modeling of changes in the condition and properties of clay seals during lithogenesis and with prediction of their screening properties. Models of clay structures and their changes with depth are considered, calculated plots of clay (clay rock) porosity and strength values for different lithogenesis stages are given, and the possibility in principle of predicting the condition and properties of clay screens is shown.

V. I. Osipov

Part I

FORMATION OF CLAY SEAL
PROPERTIES: THEORETICAL
FUNDAMENTALS

1

Composition of Clay Sediments and Their Structure Formation in Sedimentogenesis

1.1. COMPOSITION OF CLAY SEDIMENTS

Clay sediments that form in sedimentary basins represent a heterogeneous, thermodynamically unbalanced organic-mineral system which is a favorable medium for the development in it of various biochemical, physicochemical and physical-mechanical processes.

The solid part of the sediment is formed of mineral particles and organic matter that settle down to the basin bottom from aqueous suspensions. The basis of the mineral part is made up of allothigenic and authigenic minerals. The former are supplied from the continental areas surrounding the sedimentary basin, where weathering processes result in the mobilization and accumulation of sedimentary material. The principal rock-forming minerals among them are quartz, feldspars, micas, poorly dissoluble salts, clay minerals. The minerals of basic silicates and salts form, as a rule, the sand (0.05–2 mm) and silt (aleuritic) (0.05–0.005 mm) fractions of clay sediments, while clay minerals accumulate in the clay (<0.005 mm) fraction [2]. The clay (pelitic) fraction is sometimes considered as composed of particles <0.002 or even <0.001 mm in size [3].

The formation of authigenic minerals proceeds directly in the sedimentary basin medium at the expense of numerous chemical elements transported from weathering crusts. In a fluid medium, they form colloidal substances in the form of hydroxides of silicon, iron, aluminum, manganese; authigenic minerals, such as kaolinite, montmorillonite, hydromica (1M), chlorite, mixed-layer phases, may be synthesized on this basis, when alkaline metals are present in the fluid medium in sufficient quantities. Concurrently, carbonate compounds (calcite, gypsum, siderite), and in the lack of oxygen – ferrous compounds, may form chemically.

Along with the principal rock-forming minerals, sediments contain a considerable amount of organic matter – large and small vegetable detritus and skeletons of various organisms. In the near-shore environments, mainly vegetable organic matter accumulates: remains of plants brought from the land. In the deep-water part, organic matter is represented by the products of decomposition of animal and vegetable organisms inhabiting the sea: diatoms, coccolithophorids, foraminifers, radiolarians, and others.

The presence of organic matter favors the development of bacteria (sulfate-reducing, methane bacteria, etc.) which influence its decomposition. Gaseous products formed under aerobic conditions are CO_2, and methane, and under anaerobic conditions – CO_2 methane, hydrogen. Lignin turns into humic acids, albumen give ammonia and amino acids.

Clay minerals, colloidal hydroxides, and humified organic matter own extremely developed unit surface and create in the sediment a huge reserve of free surface energy in the form of uncompensated valence bonds and surface molecular forces. This causes the high activity of the colloidal and clay components of sediments in the process of structure formation and ion exchange with external solution.

The presence of clay minerals and colloids causes the development of electrokinetic phenomena in sediments. When interacting with a liquid phase, solid mineral substances adsorb on their surface ions of salts and form an electrical double layer. There are several possible mechanisms of electrical double layer formation. In clay minerals, its formation is connected with the presence of an excessive negative potential arising from heterovalent isomorphic substitutions, which is compensated by cations-compensators forming an adsorbed cation layer on the surface of minerals (Fig. 1.1, a). In a fluid medium, part of cations of the adsorbed layer dissociate and form a diffuse layer of ions (Fig. 1.1, b). The diffuse layer may be tens of nanometers thick, its thickness amounting to 100 nm and more in some cases. A change in the diffuse layer thickness causes a corresponding change in the distance from the clay particle surface, at which its electrostatic field – the electrokinetic or ξ-potential – operates.

The presence of the electrostatic field is a powerful stabilizing factor for clay minerals and colloids that allows them under certain conditions to preserve their stability and not to participate in the processes of aggregation and coagulation. Abundant physicochemical literature is devoted to the theory of

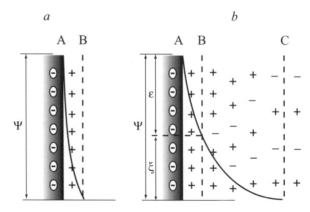

Fig. 1.1. The structure of an electrical double layer on the surface of a clay particle in the air (a) and aqueous (b) medium: ψ – thermokinetic potential, ξ – electrokinetic potential, ε – adsorption potential, AB – adsorption layer of ions, BC – diffuse layer of ions.

mineral dispersion and colloid stability [4–6]. The principal positions of this theory will be considered by us in the following paragraphs.

An important fact for the structure formation in clay sediments is that clay minerals are capable of changing the sign of potential on particle edges with a change in pH of the medium. Thus, in an acid medium, the surface of clay particles proves to be heteropotential: the basal surfaces have a negative potential and the edges have a positive potential [3, 7]. In an alkaline medium, the basal surfaces of minerals and their edges bear the same (negative) potential. This is connected with the fact that aluminum and iron hydroxides, which make part of the crystalline structure of clay minerals, show amphoteric properties and, depending on the character of the medium, may have an excessive negative or positive charge: the former is characteristic of the alkaline and the latter of the acid conditions. Organic (humic) colloids are also capable of changing their charge. Having usually a negative potential, they may, however, interact with cations of strong bases (Ca^{2+}, Mg^{2+}, Na^+, K^+) in the liquid phase to form organo-metallic compounds (humates) that bear a positive charge.

Along with the solid mineral and organic compounds, the liquid phase occupies a considerable part of the sediment volume. Water contained in sediments may be in different physical states. Free, physically bound, and chemically bound water is distinguished depending on the character of its connection with minerals [3, 8, 9].

The free water fills the sediment pores and is in no way affected by the mineral framework of the sediment. It is in the gravity field and moves in sediments in accordance with the laws of hydraulics.

The physically bound water, unlike the free water, is subject to the action of the force fields (molecular and electrostatic) of minerals. It forms a thin layer around solid particles and so is often termed the film water. Depending on the energy of its interaction with the mineral surface, the physically bound water is subdivided, in its turn, into the tightly bound (adsorbed or adsorptionally bound) and loosely bound (osmotic and capillary) water [3, 8–10].

The tightly bound water forms in the hydration of the adsorbed ion layer on the external surfaces and in the interlayer space of clay minerals; it also forms as a result of water molecule adsorption on the coordinately unsaturated atoms of the edges and on the structural OH-groups of the basal faces in the crystal lattices of clay minerals, with the coordination, hydrogen, ion–dipole, dipole–dipole, and other bonds being formed. The binding energy between the water molecules and particle surface varies from 130 kJ/mol (for the most energetically active adsorption centers) to 40 kJ/mol and less (bonds between water molecules inside the film of adsorptionally bound water) [3, 10].

Experimental researches conducted by D. Green-Kelly and B.V. Deryagin [11] have established that molecules of tightly bound water are oriented perpendicularly to the mineral surface and form a multilayer film of oriented water molecules (Fig. 1.2). However, cations of the adsorption layer cause some disturbance in the orientation of water molecules. The thickness of the formed structured hydrate cover grows with the increase in the wetting ability of a mineral and attains 1.5–6.0 nm. With temperature growth, the adsorptionally-bound

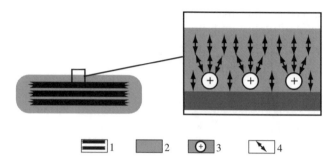

Fig. 1.2. Tightly bound (adsorbed, or adsorptionally bound) water on the surface of a clay particle: 1 – clay particle, 2 – adsorptionally bound water, 3 – exchange cations, 4 – water molecules.

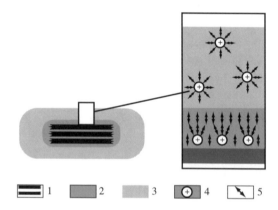

Fig. 1.3. Loosely bound (osmotic) water on the surface of a clay particles: 1 – clay particle, 2 – adsorptionally bound water, 3 – osmotic bound water, 4 – exchange cations, 5 – water molecules.

water film diminishes in thickness and is completely destroyed at 65–70 °C [10, 12, 13].

The particular structured state of the adsorptionally bound water causes its specific thermodynamic and physical–mechanical properties differing from those of free water. Thus, for example, the ultimate shearing strength of the adsorptionally-bound water film is 9.5–13 Pa, while that of the free water is 10^{-3} Pa [14]. Water in such films has lower values of dielectric permittivity, dissolving power, temperature of freezing, and enhanced heat conductivity.

The loosely bound (osmotic) water forms due to osmotic forces and makes up the external hydrate envelope of minerals (Fig. 1.3). Its thickness is commensurable with Debye's radius of ionic atmosphere ($1/\kappa$) and attains 100 nm or more on silicate surfaces. The osmotic water is bound with the diffuse layer cations and is retained by them near the mineral surface without being directly influenced by the latter. In its structural features and properties, it differs little from the volume phase. However, the osmotic water does not participate in the filtration mass transfer and its removal (squeezing-out) from the clay is possible only under significant pressure. So, the presence of this category of water exerts a great influence on the structure formation processes, filtration and physical–mechanical properties of clay sediments.

Ranked with the loosely bound category is also the capillary water formed in fine pores and capillaries at the phase boundary in the presence of a considerable content of occluded gas in sediments. In its energy state and properties, it is similar to the osmotic water, though the mechanisms of formation for these two varieties of loosely bound water are quite different.

The total content of the loosely bound water in sediment depends on its mineral composition, structure, physicochemical and pressure–temperature factors.

The third category of water is the chemically combined water, which is incorporated in the crystal lattice of clay minerals. Its removal occurs at temperatures of 500–700 °C and is accompanied by the complete disintegration of minerals.

The composition of the forming sediments plays an important role in the formation of clay seals. The processes of structure formation are, however, of no less importance, as they are sensible indicators of all the changes occurring in clay systems and determine the physical, mechanical, and filtration properties of clays at each stage of their transformations. So, the problems of structure formation in clay sediments are, along with the examination of clay composition, of great importance in studying the clay seals of oil and gas fields.

1.2. STRUCTURE OF CLAY SEDIMENTS

1.2.1. THE CONCEPT OF STRUCTURE

The term «structure» is a systems concept and is applicable for different levels of the spatial organization of substance in nature. If one traces the surrounding universe from the elemental (atomic) to the global and planetary levels, a systems series of the structure concepts can be revealed for the natural solids that surround us: atomic structure, crystal structure, rock structure, the structure of a sedimentary sequence (facies, formations) as a natural multi-rock body, the structure of the Earth's crust, the structure of the Earth, etc.

Depending on the problems to be solved, a study of the structure of natural geological bodies is conducted at different structural levels. Thus, crystallographers and mineralogists study the structure mainly at the atomic and mineral levels, petrographers, lithologists, engineering geologists, hydrogeologists – at the rock and facies (formation) levels, geologists dealing with tectonics and geophysicists – at the facies and higher levels.

The isolating properties of clays are determined by the peculiarities of the structure of both the clays proper and the sequences formed by them. So, when studying the clays as the seals for oil and gas pools, their structure should be examined at least at two levels: rock and facies. This statement is one of the fundamental ideas determining the arrangement of the present work.

The structure of clays at the rock level is characterized by the structure of the solid (consisting of mineral and organic substances) framework of clays, the porosity, and the presence of micro-heterogeneities (microbedding and microfracturing). Because of the high dispersion of clay minerals, the structure of clays is studied in laboratory with the use of high-resolution optical and scanning electron microscopes. The structural images obtained at high magnifications allow the examination of minute details of the structure, so they are often called the microstructure of clays.

The structure of clay sequences is characterized by the mode of their occurrence, variations in the lithologic composition, vertical and horizontal heterogeneity, presence of fractures. The clay sequence structure at the facies level is usually studied in situ with the use of geophysical, drilling, and other methods of investigation.

In the present work, the attention will be focused on the formation of clay structure and properties at the rock level, as well as on the characteristic features of the composition, structure and properties of clay sequences, which are determined by their depositional environment.

Before turning to the consideration of structure formation in clays, more accurate definition should be given to the concept of the structure proper, as we use it at the rock level to characterize the structure of a rock. In describing the rock structure, two approaches have been formed in geology: morphometric and energy. The former is most commonly used, wherein the concept of structure is based on the morphometric features of structural elements and their combination in the rock. In petrography, there are a number of rock structure definitions based on this principle [2, 15, 16].

In the energy approach, the concept of structure is based on such characteristics as the mode of interaction of structural elements and the energy of the structure as a whole. The energy principle has found its most extensive application in the physicochemical mechanics of fine grained systems – a new scientific trend developed by P.A. Rebinder [17] and his disciples. In the physicochemical mechanics, the structures of various natural and artificial systems are subdivided into several types according to the character of contact interactions: the coagulation, the crystallization, and the condensation structures.

Study of the energy characteristics of rock structures did not lie within the range of problems to be solved in petrography. A different situation has formed in engineering geology, geophysics, petroleum geology, where studying the physical and physical–mechanical properties of clays and clay rocks is impossible without the knowledge of their structural bonds, i.e. the character of interaction between structural elements and the conditions controlling the formation and alteration of these interactions under the influence of external and internal factors. So, when applied to the clay seals, the concept of structure cannot be considered without including the energy characteristics in it.

It has turned out historically that one more very important characteristic of the rock structure – the spatial arrangement of structural elements – is beyond the petrographical concept of structure. In petrography, these rock features are termed texture (fabric). Terminologically, the concepts of structure and texture are close to each other. Therefore, they are often substituted for one another. Thus, for example, in the literature in English they are given the opposite meaning in comparison with that in Russian and German scientific literature: the structure is called «texture», while the concept of structure, on the contrary, reflects the textural characteristics. It should be reminded that when one speaks of the atomic structure, of the mineral structure, the geometric aspect of these structures is meant first of all. It is difficult to understand then, why in discussing the rock structure its geometric aspect is forgotten. It is obvious that a broader understanding of the term «structure» should be accepted, in which its geometric features (texture), as well as morphometric and energy features are included as its partial characteristics.

So, the interpretation of the concept of rock structure, as applied to the clay seals, should include three groups of characteristics: (1) morphometric (the size, shape, character of the surface of structural elements and their quantitative proportion); (2) geometric (spatial composition of the structure); (3) energy (type of structural bonds). Taking into account the aforesaid, the concept of rock structure may be formulated as follows [18]: *structure is a spatial organization of the entire substance of the rock described by the aggregate of morphometric, geometric and energy characteristics, which, in their turn, are determined by the composition, quantitative proportion and interaction of the rock components.*

1.2.2. MORPHOMETRIC FEATURES OF THE STRUCTURE

The morphometric structural features are determined by solid (mineral and organic) structural elements composing the structural framework and by pores filled with fluid (liquid and gaseous) components of the rock. The combination of solid elements with pores in a volume characterizes the external aspect of the structure. Examination of the morphometric features of individual elements is of great importance in studying the rock structure.

Solid structural elements. The solid structural elements are understood as the elementary mineral particles, grains and their associations that form the structural framework of clays and clay rocks. The following structural elements are distinguished in clay deposits: primary clay particles, ultramicroaggregates, microaggregates, aggregates, grains, as well as inclusions of microfauna and microflora remains, microcrystals of salts and ore minerals.

The primary clay particles are represented by microcrystals of clay minerals which are resistant to mechanical and physicochemical effects. The particles of clay minerals have the shape of thin isometric or elongated plates, scales, needles, tubes (Fig.1.4). The size of clay particles is largely determined by their mineral composition and varies from tens of nanometers (montmorillonite) to several micrometers (kaolinite). The peculiarities of clay mineral particles are described in more detail in the works by L.G. Rekshinskaya [19], H. Beutelspacher and H. Marel [20].

Individual particles in clays are united, as a rule, into *ultramicroaggregates* and *microaggregates*. The former are the associations of several particles interacting usually according to the face-to-face type and having a platy or leaf-like shape (Fig. 1.5, *a*; shown by arrows). The ultramicroaggregates are most characteristic of smectites and mixed-layer minerals, for which the concept of a primary

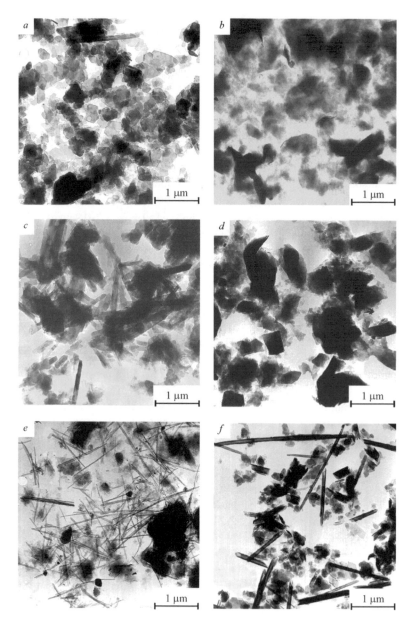

Fig. 1.4. Morphology of clay particles: a – kaolinite; b – montmorillonite; c – illite; d – mixed-layer mineral; e – palygorskite; f – halloysite.

particle is indefinite, as it means a particle having a thickness of one or several elementary layers of crystalline structure. In natural clay deposits, these minerals are encountered in the form of rather dense accumulations of clay material, which it is more correctly to consider as ultramicroaggregates rather than primary particles. Under certain conditions, ultramicroaggregates are also formed by other clay minerals (kaolinite, hydromicas). The length of ultramicroaggregates varies from fractions to several micrometers. An important characteristic of ultramicroaggregates is their rather high strength. Ultramicroaggregates are not destroyed (or are partially destroyed) during the preparation of clays for a granu-lometric analysis with the use of special mechanical and physicochemical meth-ods. Therefore, data obtained from a granulometric analysis of, for example, montmorillonite clays reflect their ultramicroaggregate composition rather than the ultimately dispersed state.

Microaggregates are the associations of clay particles and ultramicroaggre-gates not disintegrating in water in the absence of physicochemical dispersers and mechanical effects. Microaggregates are the main structural elements of natural clays. Depending on the mineral composition, shape, and size distribu-tion of primary clay particles and ultramicroaggregates, as well as on the condi-tions of their sedimentation, the microaggregates may have various framework. Their size (maximal measure) varies from several to tens of micrometers.

Most common are the microaggregates, in which clay particles and ultra-microaggregates contact along their basal planes or according to the face-to-edge type but with low angles of particle inclination, which results in the formation of anisometric structural elements of a lamellar, elongated and slightly curved shape (Fig. 1.5, b).

Kaolinites often form domain-like microaggregates consisting of axially oriented primary particles and ultramicroaggregates (Fig. 1.5, d). In some cases, isometric microaggregates are encountered in clays of polymineral com-position (Fig. 1.5, c).

A thin leaf-like shape is characteristic of the microaggregates of montmo-rillonite and mixed-layer minerals. The boundaries between microaggregates are difficult to trace, as one microaggregate gradually passes into another.

The rich in iron smectites (nontronite) and hydromicas (glauconite) have quite a specific shape of microaggregates. Both minerals form microaggre-gates of a globular shape, few to tens of micrometers in diameter, consisting of thin curved leaflets of nontronite and platy particles of glauconite.

Under certain conditions, clay microaggregates and primary silt grains form more complicated structural elements – the clay (Fig. 1.5, e) and clay-silt *aggregates* (Fig. 1.5, f) consisting of several clay microaggregates or associa-tions of clay microaggregates with silt and fine sand grains. The aggregates are, as rule, less stable in water than microaggregates.

The size of aggregates varies from tens to hundreds of micrometers, depending on the mineral composition and size distribution of clay particles, the presence in clay deposits of iron hydroxide, organic matter, and carbonates. Unlike the microaggregates, the aggregates are usually isometric. In natural clays, aggregates are not always encountered, and the boundaries between them, when examined by the aid of optical or scanning electron microscope, are rather difficult to define. The aggregates are best of all differentiated in clay deposits of eluvial and hydrothermal origin, where their outlines are emphasized by the boundaries of altered original grains of a bedrock (Fig. 1.5, e).

Fig. 1.5. Solid structure elements in clays: a – ultramicroaggregate; $b-d$ – microaggregates; e, f – aggregates; g, h – grains; i – microfauna; j – inclusion of pyrite; k – humic acid.

In this case, the presence of aggregates is determined from microfractures or thin oriented layers of clay particles at their boundaries.

The orientation of microaggregates inside the aggregates may be quite different, but the aggregates with inoriented or poorly oriented arrangement of microaggregates are prevalent. As a particular case, the aggregates with high orientation of the whole clay material are distinguished, which have been named blocks or microblocks.

The primary mineral grains in clays are represented by fragments of quartz, mica, feldspar and calcite, among which the grains of quartz composition are studied best of all. Isometric grains with different degrees of rounding and preservation are dominant (Fig. 1.5, *g*, *h*). The size of grains varies from 0.005 to 2 mm. A characteristic feature of primary mineral grains is the presence on them of «coats» of fine grained material (Fig. 1.5, *h*) represented by the particles of clay minerals, iron hydroxides, carbonates, organic-mineral compounds. The presence of the «coats» exerts a considerable influence on the contact interactions of these grains and their behavior in clays.

As inclusions, particularly in young clay sediments of marine and lacustrine origin, the remains of microflora and microfauna of various degrees of preservation are often encountered; they are usually represented by carbonate skeletons of coccolithophorids, fragments of siliceous shells of diatoms and radiolarians, spores and pollen of various plants, bacteria (Fig. 1.5, *i*). Besides, inclusions of various salts are observed in clays of marine origin and clay deposits of arid zones. These are most often the colloform accumulations of fine grained crystals of chemical calcite or the threadlike microcrystals of authigenic calcite [21].

In marine clay deposits, inclusions of pyrite (Fig. 1.5, *j*), jaipurite and other ore minerals are not rare. At last, many clays contain inclusions of organic matter in the form of little-decomposed organic remains, humic acids, and other high-molecular compounds. Decomposed organic matter actively interacts with clay minerals to form on their surface peculiar organic-mineral complexes which are well discerned through a scanning electron microscope owing to the globular aggregates of humic acids (Fig. 1.5, *k*).

Pore space.   The other component of the clay structure is pores that are formed as a result of loose arrangement of solid structural elements. The total porosity value, as well as the size and shape of pores depend on the morphometric features of solid structural elements, degree of their aggregation and compaction, character of spatial arrangement.

Under real conditions, owing to the polydispersion, anisometric shape, and small sizes of solid structural elements, the porosity of young sediments may be as high as 85–90%. At the same time, it is 20–30% in dense lithified clays and does not exceed 4–10% in argillites and siltstones. It should also be noted that the total void space value, especially in lithified clays and clay rocks, may be greatly influenced by fractures and microfractures.

Besides the *total* porosity, the open, closed and effective porosity is often distinguished in studying the clays and clay rocks [3]. The *open* porosity characterizes the total volume of interconnected pores, while the *closed* porosity characterizes the volume of isolated pores. The *effective* or *active* porosity is determined by the volume of interconnected pore space minus the volume of bound (adsorption, osmotic and capillary) water. The bound water is retained on the particle surface by molecular forces and cannot be extracted from clay in the process of free migration of pore solutions. Therefore, the effective porosity value depends on the unit surface of the clay, the viscosity of pore solutions and the pressure gradient determining the movement of the fluid.

The pore size in clays and clay rocks varies within a wide range from hundredths to tens of micrometers. Among the existing classifications of pores according to their sizes, the most well-grounded is that by E.M. Sergeev [3] worked out for the fine grained deposits and subdividing the pores depending on their genetic type and the character of water movement in them. According to this classification, the ultracapillary pores (less than 0.1 μm), micropores (0.1–10 μm), mesopores (10–1000 μm) and macropores (more than 1000 μm) are differentiated.

This classification has been taken as a basis by the authors. However, it has been found in morphometric investigations that it needs some detailing. This concerns first of all the subdividing of micropores into the fine (0.1–1 μm), small (1–10 μm), and large (10–100 μm) micropores. Besides, the value of 100 μm instead of 1000 μm has been adopted as the lower limit of micropores. So, the following pore grades are used in the subsequent descriptions: *ultramicropores* (<0.1 μm), micropores (0.1–100 μm), and *macropores* (>100μm).

Another important characteristic of pores is their shape. The following kinds of pores may be distinguished depending on their morphology [21]:

1. *isometric*, when the ratio of the linear dimensions of two most differing pore sections does not exceed 1.5. The pore size in this case is characterized by the diameter of a circumscribing circle;
2. *anisometric*, when the ratio of the linear dimensions of two most differing sections ranges from 1.5 to 10. In describing the size in this case, the length and the width of the pore are indicated (in some cases, only the average width may be indicated);
3. *fissure-like*, for which the ratio of the linear dimensions of two most differing sections exceeds 10. The size of such pores is characterized only by their width.

The presence in clays of structural elements of different sizes determines the existence of several kinds of pores. According to this characteristic, the following pores are distinguished: interparticle, interultramicroaggregate, intermicroaggregate, interaggregate, intergranular, intermicroaggregate-granular, intragranular and biogenetic pores [21].

The *interparticle* pores arise inside the ultramicroaggregates and microaggregates as a result of loose compaction of primary particles (Fig. 1.6, *a*). The sizes and shapes of these pores are determined by the morphology of interacting particles, their sizes, and the character of contacts. Inside the ultramicroaggregates, the interparticle pores have a fissure-like shape with maximal width of no more than 0.1 μm (ultrapores). In the absence of ultramicroaggregates

Fig. 1.6. Pore types in clays: *a* – interparticle; *b* – interultramicroaggregate; *c* – intermicroaggregate; *d* – interaggregate; *e* – intergranular; *f* – intermicroaggregate-granular; *g* – intragranular; *h* – pore of biogenic origin.

in clays, anisometric pores up to $0.5\,\mu m$ in width are encountered in micro-aggregates; they may also be called the interparticle, or intramicroaggregate pores.

The *interultramicroaggregate* pores are formed between ultramicroaggre-gates and may have quite different configurations depending on the spatial arrangement of the ultramicroaggregates (Fig.1.6, *b*). The fissure-like and ani-sometric pores are prevalent; the width of the former is no more than $0.3\,\mu m$, while that of the latter attains $1\,\mu m$.

The *intermicroaggregate* pores arise between microaggregates and are characterized by a great variety of sizes and shapes (Fig.1.6, *c*). The morphol-ogy of these pores depends considerably on the degree of clay lithification. Open intermicroaggregate micropores up to $15\,\mu m$ in diameter are character-istic of young weakly lithified sediments. In the clays of a medium lithification degree, isometric intermicroaggregate micropores no more than $5\,\mu m$ in diam-eter are prevalent, and in the clays of a high lithification degree – narrow fissure-like micropores up to $1.5\,\mu m$ in width.

The *interaggregate* pores in clays of sedimentary origin are seldom encountered. They are characteristic of some kinds of eluvial clays and are widespread in loess soils (Fig. 1.6, *d*). The pores are usually isometric, more rarely anisometric, their size varies from few to tens of micrometers.

The *intergranular* pores are encountered in clays with high contents of sand and silt grains. The isometric and anisometric micropores formed between the grains are from few to tens of micrometers in size (Fig. 1.6, *e*).

A certain combination of clay material and clastic grains in clays results in the formation of micropores having a more complicated structure – the *intermicroaggregate-granular* pores (Fig. 1.6, *f*), mainly isometric and aniso-metric, from few to $20\,\mu m$ in size.

In clays containing carbonate clastic grains and inclusions of microflora and microfauna, the *intragranular* pores (Fig. 1.6, *g*) are encountered, as well as the *biogenetic* pores (Fig. 1.6, *h*). The former have an isometric shape with a diameter up to several micrometers and are most often formed in carbonate grains as a result of weathering. The biogenetic pores are usually rounded in shape and fractions to tens of micrometers in size.

1.2.3. GEOMETRIC FEATURES OF THE STRUCTURE

The size and shape of solid structural elements and pores determine the mor-phometric features of the structure, while their spatial orientation determines its geometric features, or the fabric.

When the structure of clays is studied at the microlevel, its geometric features are characterized by an orderly arrangement of particles, microaggregates, grains and pores having anisometric shapes. The effect of the spatial orientation may also be enhanced owing to the arrangement of isometric micropores in one plane.

The degree or index, of orientation K_a characterizes the degree of orderly arrangement of structural elements in the orientation plane and is estimated by the ratio of the number of structural elements lying in the plane (or direction) of orientation to the total number of particles and aggregates or pores analyzed.

According to the degree of structural elements orientation, all clay deposits may be subdivided into the *poorly-oriented* (K_a = 0–7.0%), *medium-oriented* (K_a = 7.1–22.0%), and *highly-oriented* (K_a = 22.1–78.0%) clays and clay rocks [23].

With an increase in the degree of structural element orientation, a growth of anisotropy is observed in the permeability, as well as in the strength and deformation properties of clay rocks.

1.3.   THE THEORY OF STRUCTURE FORMATION IN CLAY DISPERSIONS

General regularities in the stability and coagulation of dispersed colloidal systems have been first formulated by B.V. Deryagin and L.D. Landau [5], E. Verwey and T. Overbeek [24]. The theory of stability of colloidal systems founded by them has been called DLVO (Deryagin–Landau–Verwey–Overbeek). The basis of the DLVO theory is formed by the concept of the balance of the repulsion and attraction forces operating between colloidal particles. The first determine the stabilization of clay dispersions, while the second determine their aggregation and coagulation.

1.3.1.   THE FORCES OF REPULSION

When particles in an aqueous medium draw together under the influence of thermal agitation or under mechanical effects, their direct sticking together is prevented by the forces of repulsion. These forces may be of the electrostatic and structural nature. The former are connected with the overlapping of electrical double layers (EDL) of the particles, and the latter – with the interaction of adsorption water films. The nature of both types of interaction is considered below.

One of the most critical factors determining the stability of fine grained systems is the electrostatic interaction of diffuse layer ions of EDL arising on the surface of particles in their hydration. Beyond the EDL, the intensity of the electrostatic field of particles equals zero, so the forces of repulsion arise only when the diffuse layers of particles overlap. The difference of the energy values corresponding to the not overlapped and overlapped zones of EDL is the *electrostatic component of the wedging pressure* Π_e, the calculation of which, based on the Gouy–Chapman's theory of electrical double layer and Debye–Huckel's theory of strong electrolytes, has been done by B.V. Deryagin in 1937 [6], and later by a number of other researchers.

For a common case, an equation for the electrostatic component of the wedging pressure arising between two plane surfaces has the following form:

$$\Pi_e = 64n_0kT\gamma^2 e^{-\kappa h} \tag{1.1}$$

where n_0 is the concentration of the external solution; $\gamma = th[Ze\varphi_0/(4kT)]$; Z – valence of the counter-ion; e – electron charge; φ_0 – surface potential; k – Boltzmann constant (1.38×10^{-23} J/K); T – temperature, K; κ – effective thickness of the diffuse layer; h – distance between the particles.

For weakly charged surfaces, we have:

$$\Pi_e \approx \frac{4Z^2 e^2 \varphi_0^2 n_0}{kT} e^{-\kappa h} \tag{1.2}$$

It is seen from Equation (1.2) that for weakly charged surfaces, Π_e is proportional to the square of the surface potential φ_0.

For strongly charged surfaces, the wedging pressure may be estimated as follows:

$$\Pi_e \approx 64 n_0 kT e^{-\kappa h} \tag{1.3}$$

In this case, Π_e does not depend on the surface potential, as the surface is screened by the counter-ions.

Equations (1.2)–(1.3) refer to the interaction of two particles having plane surfaces. In calculating distant interactions for spherical particles, the following solution is often used:

$$\Pi_e = \frac{8(kT)^2 \varepsilon r}{e^2 Z^2} \gamma^2 e^{-\kappa h} \tag{1.4}$$

where ε = dielectric constant of the medium, r = particle radius.

As is seen from Equation (1.4), the electrostatic component of the wedging pressure for spherical particles is in proportion to their radius r.

Another approximate solution has been derived for the spherical particles with a low value of potential φ_0 provided that their radius is significantly larger than the diffuse layer thickness ($\kappa r \gg 1$):

$$\Pi_e = \frac{\varepsilon r \varphi_0^2}{2} \ln\left[1 + e^{-\kappa h}\right] \tag{1.5}$$

When $\kappa h \ll 1$ (small particles in weak electrolytic solutions), the following dependence is used:

$$\Pi_e = \frac{\varepsilon r \varphi_0^2}{h + 2r} e^{-\kappa h} \tag{1.6}$$

The other component of the wedging pressure operating between particles arises during the interaction of structured adsorption water films. It manifests itself when particles draw so close together (on the order of 1.5–6.0 nm) that their hydrate adsorption layers become overlapped.

As has been already noted, thin films of water molecules adsorbed on the mineral surface are in a particular structural state and their properties differ from those of free water. When these films overlap, the peculiar structure of their peripheral parts is destroyed at the contacts, so that part of the adsorbed water passes into the volume phase. This leads to the change in the free energy of the hydrate film and to the manifestation of the wedging pressure. The presence of such forces has first been established by B.V. Deryagin and E.V. Obukhov in 1935 [25]. Later, B.V. Deryagin and N.V. Churaev [26] have termed them the *structural forces*, or *the structural component of the wedging pressure* (Π_s). A characteristic feature of Π_s is the quick growth of its value with the decrease in the distance between particles. The calculations show that at short distances between particles, the structural wedging pressure exceeds by far the electrostatic wedging pressure.

The existence of the structural wedging pressure and its dependence on the adsorption film thickness have been experimentally proven by B.V. Deryagin and Z.M. Zorin [27], B.V. Deryagin and N.V. Churaev [26], with the wettable surface of quartz and glass taken as an example. Later, similar data have been obtained for the plates of mica [28, 29]. Australian physicochemists J. Israelachvili and R. Pashley [30] connect the wedging effect with the hydration of cations and call it the hydration effect.

Calculation of $\Pi_s(h)$ has been done for the first time by S. Marčelja and N. Radič [31] on the basis of an empirical dependence obtained by them:

$$\Pi_s(h) = K \exp(-h/l) \qquad (1.7)$$

where K and l are empirical constants, whose values for different conditions of measurements have been tabulated by B.V. Deryagin and N.B. Churaev [12]. The obtained results have been in sufficient accord with the experimental data.

1.3.2. THE FORCES OF ATTRACTION

The forces of attraction include the molecular, or van der Waals forces operating between individual molecules and atoms. They are much weaker than the forces of chemical nature, unsaturated, and may be considered in the first approximation as additive. Thanks to this property, such forces arise in the drawing together of both colloidal particles and macroscopic bodies. Having a large radius of action, the molecular forces play an important part in the interaction of fine grained particles, where they exceed the gravity forces. When the interaction of fine grained particles is considered, they are often called the *molecular component of the wedging pressure*, Π_m [12, 32].

There are two theories for the calculation of molecular forces: the microscopical theory worked out by F. London [33] and H. Hamaker [34] and the macroscopical theory developed by E.M. Lifshits in his works [35, 36]. According to the first of them, the energy of molecular attraction between two spherical particles with similar radii r is determined from the following expression:

$$U_m = -\frac{A}{6}\left(\frac{2r}{R^2 - 4r^2} + \frac{2r^2}{R^2} + \ln\frac{R^2 - 4r^2}{R^2}\right)$$

(1.8)

where A is the van der Waals–Hamaker constant (for water, $A = 1.12 \times 10^{-20}$ J); R = distance between the particle centers.

For the $r \gg h$ condition, where h is distance (clearance) between two spherical particles, we have:

$$U_m = -Ar/(12h)$$

(1.9)

and for a spherical particle and a plane:

$$U_m = -Ar/(6h)$$

(1.10)

For the energy of molecular attraction between two plane-parallel plates with thickness δ, E. Verwey and T. Overbeek [24] have obtained the following expression:

$$U_m = -\frac{A}{48\pi}\left[\frac{1}{(h/2)^2} + \frac{1}{(h/2 + \delta)^2} + \frac{2}{(h/2 + \delta/2)^2}\right]$$

(1.11)

With some assumptions taken, Equation (1.11) may be simplified. Thus, for thick plates with $\delta \gg h$, we have:

$$U_m = -A/(12\pi h^2)$$

(1.12)

The Π_m value of the wedging pressure is found in this case from the expression:

$$\Pi_m = -A/(6\pi h^3)$$

(1.13)

The negative sign of the wedging pressure evidences for the operation of forces causing the particles to draw together under the influence of molecular attraction.

For thin plates, when distance h is much greater than their thickness, i.e. $\delta \ll h$, the energy of interaction is determined as follows:

$$U_m = -A\delta^2/(2\pi h^3)$$

(1.14)

Equations (1.8)–(1.14) deal with the interaction in vacuum and are applicable only for short distances, when the effect of electromagnetic delay may be neglected. This effect is taken into account in the second theory of molecular forces, which has been given the name of the macroscopical theory and is based on the fluctuation electrodynamics and on the hypothesis that the interaction of bodies is realized through radiated by them or fluctuation, electromagnetic fields. In 1954, E.M. Lifshits has obtained formulae for calculating the force of attraction between two plates related to the surface unit for two eultimate cases:

1. with a small clearance value, when $h < \lambda$:

$$\Pi_m = \frac{\hbar}{8\pi^2 h^3} \int\limits_0^\infty \left(\frac{\varepsilon(i\xi)+1}{\varepsilon(i\xi)-1} \right)^2 d\xi \qquad (1.15)$$

2. with a great clearance value, when $h > \lambda$:

$$\Pi_m = \frac{\hbar c}{h^4} \frac{\pi^2}{240} \left(\frac{\varepsilon_0 - 1}{\varepsilon_0 + 1} \right)^2 \varphi(\varepsilon_0) \qquad (1.16)$$

where λ is the length of electromagnetic wave, $\hbar$ – Planck's constant related to 2π; $\varepsilon(i\xi)$ – complex dielectric permittivity, which is a function of the imaginary variable; ξ – the imaginary component of the complex frequency; c – velocity of light; $\varphi(\varepsilon_0)$ – a certain function of the statistical dielectric permittivity ε_0.

By integrating Equation (1.16), the expression is obtained for determining the energy of attraction between two plates:

$$U_m = -B/h^3 \qquad (1.17)$$

where B is the constant of molecular attraction forces (for quartz with $\varepsilon_0 = 3.6$, $B = 4.4 \times 10^{-27}$ J $\times$ cm).

Because of restricted available data on the complex dielectric permittivity values for different solid bodies, the calculation of particle interaction at great distances is usually performed on the basis of the microscopic theory, with the obtained data being corrected for electromagnetic delay. The correction factors for different conditions of measurement are given in the form of tables by H. Kruyt [37].

1.3.3. THE SUMMARY FORCES OF PARTICLE INTERACTION

The stability and the rate of aggregation of fine grained systems depend on the sign and value of the summary energy of their interaction determined by the relation of their repulsion and attraction energy. When stability is considered in terms of the modern concepts about the nature of forces operating between particles and the role of thin adsorption films in the contact clearance, all the above-described components of the wedging pressure should be taken into account [38]:

$$\Pi = \Pi_e + \Pi_m + \Pi_s \qquad (1.18)$$

where Π_e, Π_m, and Π_s are the electrostatic, molecular, and structural components of the wedging pressure, respectively.

However, because of poor development of the theory used for finding the structural component of the wedging pressure, estimation of the fine grained system stability in the framework of the DLVO theory is usually accomplished using two components – electrostatic and molecular. Numerous examples evidence that application of the DLVO theory gives satisfactory results in calculating the interactions at great and medium distances (no less than 10 nm) between particles. With lesser distances, the results are less reliable, which is obviously connected with disregarding the structural component of the wedging pressure.

According to the DLVO theory, the stability of particles is determined using the following equations:

$$\Pi = \Pi_e + \Pi_m = 64 n_0 k T \gamma^2 e^{-\kappa h} - \frac{A}{6\pi h^3} \tag{1.19}$$

$$U = \frac{64 n_0 k T \gamma^2}{\kappa} e^{-\kappa h} - \frac{A}{12\pi h^2} \tag{1.20}$$

Solutions to Equation (1.20) for different conditions show that there are three principal varieties of the summary curve describing the energy of interaction between particles as a function of distance between them (Fig. 1.7). Curve 1 corresponds to such a condition, when attraction energy prevails at any distance between particles. This is characteristic of the interaction between likely charged hydrophobic particles with low surface potential, as well as between oppositely charged particles or particles with the heteropotential character of charge distribution (for example, clay particles in acid medium). Such systems are subject to quick aggregation or coagulation and cannot exist in a stable dispersed state.

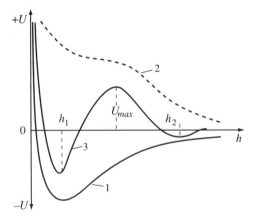

Fig. 1.7. Dependence of the total interaction energy (U) of two plates on the distance between them (h). The condition: 1 – not stabilized at any distance; 2 – stabilized at any distance; 3 – not stabilized at distances h_1 and h_2.

Curve 2 corresponds to the condition when the energy of repulsion between particles at any distance exceeds the energy of thermal agitation of the particles. Under these conditions, the probability of overcoming the force barrier is negligible, and hence, the rate of coagulation practically equals zero. Such a phenomenon is usually observed for hydrophilic or strongly charged hydrophobic particles whose surface is stabilized by organic molecules and other surface active substances enhancing the stability of particles in solutions. Fine grained systems in such a state are stable and capable of staying in the dispersed form for a long time.

Most typical of natural clay systems is the energy curve of the third kind (Fig. 1.7), which is characterized by the presence of two potential minimums at distances h_1 and h_2 (called «close» and «distant», respectively), as well as of the repulsion barrier U_{max} separating them. The molecular attraction forces prevail at short distances, causing the existence of the close potential minimum. The position and depth of this minimum are influenced by the Born's repulsion forces between the atoms of the surfaces being drawn together. The formation of the distant potential minimum at $h_2 \gg 1/\kappa$ is also connected with the molecular forces. The prevalence of the molecular forces at these distances is explained by the fact that with the growth of h, they decrease more slowly than the forces of electrostatic repulsion.

The systems having this type of the summary interaction curve are characterized by the formation of aggregates at two fixed distances between particles, h_1 and h_2, corresponding to the close and the distant potential minimum. Besides, stable hydrate films form at the contacts of particles composing the aggregates. In the first case, the film thickness is several nanometers, in the second case it is tens of nanometers. Fixation of particles in the distant potential minimum occurs under the condition that the kinetic energy of the particle thermal agitation, kT, is less than the depth of the distant potential minimum. If this condition is not met, the system is stable at high values of the energy barrier. With the lowering of the latter, the system may again lose its stability due to the fixation of particles in the close potential minimum.

The summary energy curve of the interaction of fine grained bodies changes under the effect of quite a number of factors influencing the value of the electrostatic or molecular interaction energy. This explains the great variety of structure formation processes under natural conditions.

1.3.4. REGULARITIES OF THE AGGREGATION AND COAGULATION IN CLAY DISPERSIONS

There is no agreement in opinions in geological and physicochemical literature concerning the «aggregation», «coagulation» and «flocculation» terms, which creates much difficulty in their use.

The aggregation is usually understood as the process of formation of enlarged structural elements resulting from the sticking together of primary elements that have lost their stability. It develops in diluted dispersions and does not bring about the volume structuring of the latter. The aggregates participate in sedimentogenesis independent of one another, obeying the gravity forces. The coagulation is usually considered as the process of interaction of primary particles or their associations (ultramicroaggregates, microaggregates and aggregates) in concentrated dispersions (gels) or sediments, resulting in the formation of a continuous volume structure of the solid phase. In both cases, aggregation or coagulation, the process involves diminishing of the surface energy of the system. The term «flocculation» will not be used below, as it is synonymous in its meaning to the term «aggregation».

Regularities of the aggregation and coagulation in clay dispersions depend on the character of the summary energy interaction of mineral particles. Among the main factors determining the conditions of stabilization or aggregation of fine grained systems, there are the size and shape of particles, their concentration and homogeneity in size, the value of charge and the character of its distribution, the concentration and composition of the external solution, pH of the medium, temperature, various external force fields (gravity, electric, magnetic), hydrodynamic and acoustic effects, etc. An analysis of individual factors determining the stability of fine grained particles has been accomplished by E.D. Shchukin et al. [32].

The influence of the size, shape, and concentration of particles in a volume has been analyzed most thoroughly by I.F. Efremov [39]. In studying the interaction of plates, he has revealed the following regularities:

1. at a low volume concentration of particles, thin plates have a higher energy barrier and hence, higher stability. Thick plates are characterized by the presence of deep distant potential minimum and by the aggregation in that minimum;
2. with a growth of the concentration of plates in a solution (the conditions of straitened coagulation), their pair interaction is influenced by the neighboring particles (the cooperative effect), which leads to an increase in the distant potential minimum depth and decrease in the stability of plates as a result of coagulation in the distant minimum. Under these conditions, thin plates prove to be less stable than thick ones. The conditions of straitened coagulation for thin plates set in at a lower concentration than for thick plates.

In the interaction of spherical particles, the regularities are different:

1. the stability of spherical particles, unlike that of plates, lowers with the diminishing of their radius;
2. the presence of distant potential minimum is more characteristic of large particles than of small ones, and so the aggregation processes in the distant minimum are most spread in coarsely dispersed systems, while the aggregation processes in the close minimum are most common in finely dispersed systems.

In view of the wide distribution of polydispersed systems, it is very important to study the conditions of heterocoagulation, i.e. interaction of particles having different sizes. I.F. Efremov [39] has established that, when large and small particles of flat and spherical shapes are present in the system, the probability of aggregation is higher in the collision of large particles with one another or of a large and a small particles, while the collisions of small particles have little effect because of their high energy barrier and small distant minimum depth. For example, the rate of sticking of small flat particles on the large ones is 400–500 times higher than the rate of small particle aggregation. If heterogeneous particles are evenly distributed, the probability of collisions of large and small particles is higher than that of particles having similar sizes. This results in the formation of the aggregates composed of large and small particles, wherein small particles settled on the surface of large particles block their surface forces. The polydispersed aggregates acquire stability and do not come into interaction with one another. It is obvious that development of heterocoagulation processes depends to a great extent on the optimal combination of large and small particles in the system.

It should be noted that the shape of particles influences the character of charge distribution on their surface and the degree of stabilization of different portions of the surface. With an increase in the surface curvature, the degree of its stabilization by the electrostatic forces diminishes; therefore, in natural clay systems, the probability of aggregation of spherical and flat particles is higher than the probability of aggregation of two flat particles. The same reason accounts for the growing probability of flat particle aggregation according to the face-to-edge type, when one particle's edge having an enhanced surface curvature interacts with a flat (basal) surface of the other particle. The number of such interactions grows sharply in an acid medium, when faces and edges have different charges. The clay system loses its stability completely under such conditions.

The shape of particles also influences the value of the solid phase critical concentration in a suspension, at which the coagulation occurs and a continuous volume structure forms in the sediment. When particles interact according to the face-to-edge type, this concentration is the lower, the more anisometric the particles. For fine grained spherical particles, the critical concentration value may be equal to several per cent, while for a system composed of platy particles it is several tenth or few per cent, and for tubular particles and bands – several hundredth, more seldom tenth, per cent.

The stability of fine grained systems is influenced to an extremely high degree by the concentration and composition of electrolytes in the external solution. A theoretical analysis of the influence of this factor is given in the works of I.F. Efremov [39], G. Sonntag and K. Strenge [40], E.D. Shchukin et al. [32], B.V. Deryagin et al. [41]. Adding the electrolyte causes a decrease in the electrostatic potential of the surface (due to the adsorption of counter-ions) and a contraction of the diffuse layer of EDL. In clay systems, both processes may go concurrently: on the edges of particles, the surface potential decreases as a result of the counter-ion adsorption, while on their basal surfaces it decreases due to the diffuse layer contraction. At the same time, the depth of the distant potential minimum increases and the height of the energy barrier lowers. The process is accompanied by the aggregation of particles in the distant or close potential minimum (see Fig. 1.8). A slight growth of the electrolyte concentration results in an increase of the distant potential minimum depth and in the aggregation of particles in the distant potential minimum. A further increase in the electrolyte

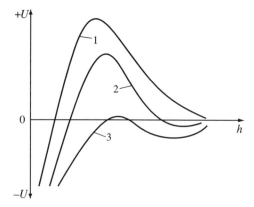

Fig. 1.8. Dependence of the summary interaction energy U of two flat particles on the distance h between them at different concentrations of electrolyte in the external solution: 1 – in the absence of electrolyte; 2 – at the low concentration of electrolyte; 3 – at the high concentration of electrolyte.

concentration causes a decrease in the energy barrier height and in the strengthening of the system at the expense of particle aggregation in the close potential minimum.

1.4.   STRUCTURAL BONDS

The processes of aggregation and coagulation are accompanied by the formation of structural bonds between structural elements, i.e. of stable attraction forces that cause the existence of the volume structure of the sediment. The structural bonds are the energy characteristic of the structure, with which many physical and mechanical properties of clays are connected.

The structural bonds are formed not all over the phase interface of structural elements, but at places of their maximal closeness – at the contacts. The contacts are the weakest points of the structure, along which its deformation and destruction occur.

The theory of contact interactions in fine grained porous bodies has been worked out in the beginning of 70s by P.A. Rebinder [17] and his disciples [42–47]. In application to clays, it has been further developed in the works of I.M. Gor'kova [48], I.G. Korobanova [49], V.I. Osipov [50, 51], V.I. Osipov and V.N. Sokolov [52, 53], V.N. Sokolov [23, 54, 55], L.I. Kul'chitskiy and O.G. Us'yarov [56] and others.

Despite the variety of mineral compositions, sizes and shapes of structural elements that make up the clay deposits, three principal types of contacts may be distinguished, which are commonly encountered in fine grained structures: coagulation, transitional (point) and phase contacts. Each type of contact is characterized by a definite mechanism of formation, by the nature of forces operating in the contact zone, by the geometry and value of contact interaction. A description of the coagulation contact is given below; the transitional and phase contacts will be considered in the next chapter.

Fig. 1.9. A scheme of the coagulation contact: 1 – clay particle; 2 – bound water.

The processes of aggregation and coagulation of clay particles in an aqueous medium result in the formation of sediments with *coagulation contacts* between structural elements. The cohesion between particles at such contacts is due to the long-range molecular and in some instances also magnetic and dipole (Coulomb's) interactions [54, 56, 57] the summary energy of which exceeds the energy of thermal agitation of the interacting particles.

Description of the conditions controlling the formation of coagulation contacts in fine grained systems is given in the works of P.A. Rebinder [17], E.D. Shchukin [42], V.V. Yaminskiy et al. [38], and many others. A characteristic feature of the coagulation contacts in clay dispersions is the presence at the contact between particles of a thin equilibrated interlayer of a liquid (Fig. 1.9), whose thickness corresponds to the free energy minimum of the system and may vary from few nanometers to 50–80 nm and more.

The coagulation contacts form under the equilibrium conditions and are completely mechanically (thermodynamically) reversible. Owing to this, clay sediments tend to show the plastic flow, i.e. to deform under constant loading without the break of continuity and, when destroyed, to restore thixotropically-reversibly their structure.

The most important role in the formation of coagulation contacts belongs to the van der Waals forces, or the forces of molecular attraction, that arise at the expense of pair interactions of neutral molecules, both having and not having the dipole moment. In the first case, the electrostatic interaction occurs due to the orientation (both of the molecules have dipoles) and induction (one of the molecules has a dipole) effect, while in the second case the electromagnetic interaction occurs due to the dispersion effect. The pair interactions of molecules may sum up additively to create a field of molecular attraction operating between any solids when they draw together at a distance measured by tens and hundreds of nanometers. With a growth of the system dispersion (unit surface), the role of molecular attraction in the formation of the volume structure increases.

The general theory of molecular forces has been set forth in the foregoing paragraph. By integrating expression (1.13), one may have an idea of the potential energy of the molecular interaction between two plates:

$$F(h) = \sum_{m}^{\infty} \Pi(h) \, dh \qquad (1.21)$$

The value $F(h)$ characterizes the unit (related to m^2) energy of cohesion of two solids and is invariant with respect to the size and shape of particles, as well as to the nature of forces operating at the contact. If the interacting particles are similar in their nature, it is more convenient to use the value $F(h)/2$ as a specific characteristic of the force field of a solid surface. B.V. Deryagin et al. [58] have succeeded in finding a strict thermodynamic correlation allowing the strength estimation through the value $F(h)$:

$$P'(h) = \pi \overline{R} F(h) \qquad (1.22)$$

where $P'(h)$ is the strength (cohesion) of an individual contact determined by the force required for the tearing of two particles from one another; $\overline{R}$ – radius of curvature of the contacting surfaces; $F(h)$ – unit free energy of interaction of two plane surfaces.

As noted above, the interaction energy of two charged particles in liquid medium was first quantitatively described by Deryagin and Landau [5], as well as Verwey and Overbeek [24] proceeding from the general concept of molecular attractive forces and ion-electrostatic repulsion of diffusive layers of likely charged cations around particles. This theory development gave birth to the idea about two possible potential minimums in the curve of particle interaction energy built as a function of the distance between particles. The first one lies at a distance measured in nanometers (h_1); and the other (h_2), tens of nanometers (Fig. 1.7, curve 3). Hence, two types of coagulation contacts were distinguished, i.e. close and distant contacts fitting particle fixation at close and distant potential minimums, respectively.

Along with the molecular attraction, the electrostatic interactions may arise between the particles separated by a liquid dispersion medium. This is caused by a rigid dipole moment that particles acquire under certain conditions [7, 59]. The dipole moment formation is connected with the recharging of clay particle edges in the acid medium and the creation of oppositely charged (heteropotential) portions of the particle surface (Fig. 1.10, a). In the alkaline medium, the edge and the basal surface of a particle acquire the like potential and the particle loses its dipole properties (Fig. 1.10, b). Investigations by N.A. Tolstoy et al. [60] have shown that the formation of the particle dipole moment may be also promoted by the unipolar orientation of adsorbed water molecules on the mineral surface.

The presence of a rigid dipole results in the rise of electrostatic attraction (Fig. 1.10, c) between particles. A calculation of such interaction performed for kaolinite shows that the force of attraction between two micron-sized particles of this mineral is $\sim 1.2 \times 10^{-9}$N [54]. Despite the small value of the forces under consideration, they may play an important role in the formation of structural bonds in young clay sediments. A confirmation to this may be the behavior of clay dispersions in various media: in a weakly acid medium they

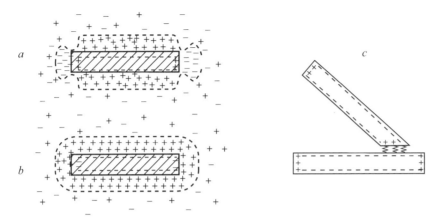

Figure 1.10. The sign of potential on the surface of clay mineral particles in the acid (*a*) and alkaline (*b*) media. A scheme of electrostatic interaction between two particles in the acid medium (*c*).

instantly coagulate, while in a weakly alkaline medium they form a stable, well dispersed suspension.

Besides the molecular and dipole interactions, the forces of magnetic nature may take part in the formation of the structural coherence of fine grained soils [61]. This is accounted for by the fact that clay particles have a magnetic moment, which leads to their interaction under the geomagnetic field conditions. Calculations show that the value of the magnetic component of structural bonds at a single contact of two clay particles does not exceed 10^{-9}–10^{-11} N. Consequently, the role of magnetic forces in the formation of the structural coherence of fine grained soils is small. The action of these forces should be probably taken into account only at the early stages of sediment formation, when they may assist in the coagulation of particles.

Estimating the force of particle cohesion in a coagulation contact, i.e. the strength of an individual contact, is of great scientific and practical interest. Application of the DLVO theory for such estimates does not give satisfactory results. This may be explained by at least two reasons. One of them is that a powerful stabilizing factor in clay systems is provided, besides the ionic–electrostatic repulsion, by the interaction of boundary hydrate layers having elastic-viscose properties. There are no quantitative methods for determining the interaction of such layers, which makes it very difficult to take them into account in the general balance of forces operating at the contact. The other reason may be a change in the sign of the ionic–electrostatic interaction between particles at short distances, when the ionic–electrostatic forces cease to be a stabilizing factor and become a factor causing the particle attraction. Besides, the DLVO theory does not take into account the possible contribution to the particle interaction of forces having the electrostatic and magnetic nature.

More accurate concepts about the strength of individual coagulation contacts are obtained based on experimental microscopic determinations.

B.V. Deryagin and I.P. Abrikosova [62] have worked out an original technique and conducted measurements of the cohesion forces between two quartz threads in various liquids. Interesting results have been obtained by J. Israelachvili and G. Adams [28] and by Y.I. Rabinovich [63]. Later, a series of original experiments has been carried out by E.D. Shchukin et al. [43], V.G. Babak et al. [64]. The direct methods for determining the strength of individual contacts are described in more detail in Chapter 8. Here, we shall only consider some of the data obtained in these experiments.

The investigation by J. Israelachvili and G. Adams [28] is the most interesting with respect to clay sediments. It has allowed the estimation of interaction between two mica plates in water and in aqueous electrolytic solutions, with the contact clearance between the plates being gradually diminished up to their direct contact. A complete potential interaction curve having a form similar to that of curve 3 (Fig. 1.7) has been measured, and the distant potential minimum at a distance $h_2 \approx 7.5$ nm and the close potential minimum at a shorter distance (h_1) have been established. The unit cohesion energy values ($-F(h)/2$) corresponding to them were equal to 4×10^{-3} mJ/m^2 and 0.01–10 mJ/m^2.

The conducted direct measurements and obtained values of ($-F(h)/2$) allow the calculations of the strength of coagulation contacts between particles of different sizes. The results of calculations according to formula (1.22) show that the coagulation contact strength in the distant potential minimum is no more than 10^{-10} N for two smooth particles having a diameter $2r \approx 1$ μm (or the same for coarser rough particles having a radius of the irregularity and surface curvature $r \approx 0.5$ μm). The coagulation contact strength in the close potential minimum is much higher and ranges from 1×10^{-9} to 5×10^{-8} N. Such a wide scattering of the $P'(h)$ values is explained by a geometric factor, on which the area of the formed contact depends: the minimal value (1×10^{-9} N) is characteristic of two micron-sized spherical particles and the maximal one is for two parallel flat particles of the same size. With an increase in the particle size, the strength of the coagulation contact significantly increases. For example, the $P'(h)$ value for two spherical particles 1 mm in diameter is $\sim 10^{-5}$ N. So, when considering the contact interactions, it is necessary to indicate the average size of the particles for which the contact strength is to be determined.

The calculations show that the tensile strength of a fine grained system with porosity of 40–50%, average particle size of 1 μm, and $P'(h) \approx 10^{-8}$ N is $\sim 10^4$ Pa. Such strength values are characteristic of many clay muds and weakly lithified clays.

1.5. FORMATION OF PRIMARY MICROSTRUCTURES IN CLAY SEDIMENTS

The processes of structure formation in clay sediments have a complex nature and develop at each stage of sedimentogenesis. Their development begins in the near-bottom part of a water basin in clay dispersions – the natural aqueous solutions, in which clay particles are transported in the suspension state – and goes on in the sediment proper.

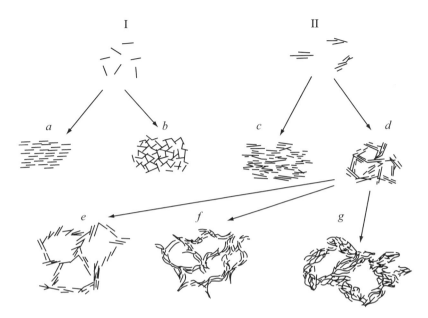

Fig. 1.11. The condition of clay matter in aqueous dispersions: I – dispersed; II – aggregated; *a–d* – models of clay sediment microstructures (*a, b* – dispersed-coagulated; *c, d* – aggregated-coagulated); *e–g* – models of the honeycomb-type aggregated-coagulated microstructure formed in clay sediments of the kaolinite, montmorillonite, and polymineral (predominantly hydromica) compositions, respectively.

In natural aqueous solutions, the clay matter is in the dispersed or aggregated state (Fig. 1.11). It should be noted that stability of individual clay particles has a dynamic rather than thermodynamic nature, i.e. their aggregation under the influence of stabilizing physicochemical factors develops extremely slowly, which suggests the possibility of the dispersed state of particles. Such a state is observed at extremely low concentrations of a solid phase in the solution, when the probability of particle collisions is insignificant, as well as in a weakly alkaline medium or under the conditions of particle stabilization by the high-molecular organic matter playing the role of surface-active compounds.

It is more probable that clay matter is present in the natural aqueous dispersions in the aggregated state. Aggregation may be quick or slow and develops as a result of partial or complete loss of stability by particles under the influence of various factors. In case of the partial loss of stability, the edges of particles prove to be their less stabilized portions, so the collision of particles results in their mutual fixation according to the face-to-edge type. When the loss of stability is complete, the aggregation occurs mainly according to the face-to-face or face-to-edge at a low angle type. In both cases, the particle fixation may be realized at the expense of the distant or the close potential minimum. In heterogeneous systems, mainly polydispersed associations form due to the predominant aggregation of small and large particles.

The size of particle associations increases up to a certain limit, at which they acquire dynamic stability and lose the capability of joining new particles. The critical size, at which the stabilization sets in, depends on specific physico-chemical conditions. The variety of the latter results in the formation of either ultramicroaggregates consisting of several particles or larger microaggregates consisting of the primary particle or ultramicroaggregate associations.

Deposition of dispersed particles, ultramicroaggregates and microaggre-gates gradually leads to the increase of their concentration in the near-bottom part of a water basin and to the development of coagulation process, i.e. to the volume structure formation. The formation of different microstructure types, whose models are shown in Figure 1.11, is theoretically possible in this process. The first two microstructures (Fig. 1.11, a, b) are formed as a result of dispersed clay matter sedimentation. In this case, the coagulation develops in the near-bottom part of a water basin or in the sediment proper, when elements with partly stabilized surface coagulate. The latter is most characteristic of the anisometric particles having a tubular or a platy shape, which have lower stabilization of edges in comparison with basal surfaces. The same effect is observed in an acid medium and possibly in a neutral medium due to the presence of positive charges on the edges. In both cases, after the suspension in the near-bottom part of a water basin reaches a certain concentration, it coagulates and forms a high-porosity spatial structural framework of the «card house» type (Fig. 1.11, b).

Under the conditions, when clay matter in the dispersed state is well stabi-lized, it is capable of settling down to the water-basin bottom and forming a suf-ficiently compact sediment with good orientation of particles along the bedding (Fig. 1.11, a). After the accumulation of sediment and, possibly, some com-paction of it, the processes of straitened coagulation in the distant potential minimum develop in it as a result of the collective interaction of particles. The microstructures formed in both cases (Fig. 1.11, a, b) may be called the dispersed-coagulated microstructures.

When the aggregated clay matter is deposited, the formation of microstruc-tures similar to the above-discussed is possible, the only difference being that they are formed not by individual particles but by ultramicroaggregates and microaggregates. In the near-bottom part of a water basin, the coagulation of ultramicroaggregates and aggregates proceeds according to the face-to-edge and edge-to-edge types, which causes the formation of a high-porous honey-comb type structure (Fig. 1.11, d). When ultramicroaggregates and micro-aggregates are well stabilized, their independent deposition and straitened coagulation occur, i.e. a well-oriented, compact sediment is formed (Fig. 1.11, c). These microstructures may be classified according to the mechanism of their formation as the aggregated-coagulated microstructures.

The above-considered models of clay sediments are purely theoretical and valid for the systems with no less than 40% of clay material, when larger clas-tic grains do not influence the development of the structure formation process. Investigations of natural clay sediments carried out by R. Push [65], P. Smart [66], V.I. Osipov and V.N. Sokolov [67, 68], V.I. Osipov et al. [69] have shown that the first three microstructures are encountered extremely seldom under natural conditions. They can be produced artificially in a laboratory or industrially with the corresponding chemical treatment of clay dispersions.

The majority of researchers, who studied natural clay sediments, used to note the prevalence in them of the aggregated-coagulated microstructure of the honeycomb type (see Fig. 1.11, d). This microstructure is encountered both in the marine and fresh-water basin environments. Depending on the depositional environment and mineral composition, the formed microstructures differ in the structure of microaggregates and in the character of their interaction in a cell. Three varieties of honeycomb microstructures are most common. The first one is characteristic mainly of kaolinite sediments and is formed by elongated microaggregates, whose structure resembles a shifted pack of cards (Fig. 1.11, e). The particles inside the microaggregate and the aggregates between themselves interact chiefly according to the face-to-face and face-to-edge at a low angle types.

The second variety of the honeycomb microstructure is peculiar to clay sediments of the montmorillonite and mixed-layer composition (with abundant swelling layers). It is composed of dense leaf-like microaggregates of clay particles having an irregular shape with slightly twisted edges (Fig. 1.11, f). The microaggregate boundaries are difficult to discern, as the microaggregates overlap one another, interacting according to the face-to-face type and forming a closed cell. Models similar to those shown in Figure 1.11, e, f have been first obtained by N. O'Brien [70] based on the study of artificial kaolinite and hydromica sediments.

The third variety of the honeycomb microstructure is most characteristic of present-day clay sediments (Fig. 1.11, g). It has been described for the first time by R. Push [71]. Such a structure is most often encountered in sediments of the hydromica and polymineral composition. Their characteristic feature is the presence of large isometric microaggregates formed by the particles interacting according to the face-to-edge at a low angle and face-to-face types.

As an example, Figure 1.12, a shows a photomontage of SEM images of the honeycomb microstructure in a present-day polymineral clay sediment from the Black Sea. As can be seen from the figure, the microstructure is formed by large flattened microaggregates contacting according to the face-to-face and face-to-edge types and making up the closed isometric intermicroaggregate pores-cells (1). The walls of the cells are composed of elongated and anisometric microaggregates (2) contacting according to the face-to-edge type (3). The length of the microaggregates varies from 3 to 7 μm and their thickness is 0.4 to 1.2 μm. Isometric microaggregates 4.5–7.0 μm in size are encountered at the joints of cells. The microaggregates (1) shown in Figure 1.12, b have a complex structure and consist of anisometric ultramicroaggregates (2) contacting according to the face-to-face and face-to-edge at a low angle types. The length of the microaggregates is 0.5–2 μm and the thickness is 0.1–0.3 μm.

Besides the ultramicroaggregates and microaggregates, rare isometric silt grains (5) covered with clay «coats» are encountered in the sample (see Fig. 1.12, a). An average size of the silt grains varies from 9 to 19 μm.

Among the solid structural elements (Fig. 1.12, a), there are rather numerous inclusions of microfauna: skeletons of coccolithophorids having different

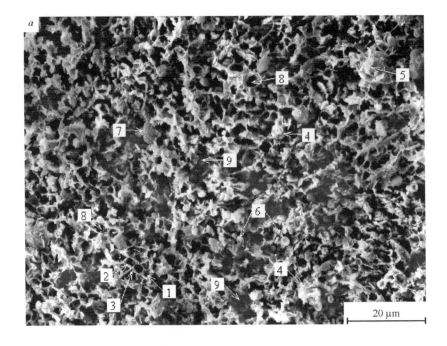

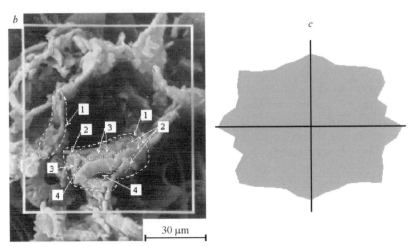

Fig. 1.12. A general view (*a*) of the honeycomb microstructure of a present-day clay sediment from the Black Sea, its fragment (*b*), and the orientation rose of structural elements (*c*).

degrees of preservation, rare fragments of diatom shells (7). The coccolithophorids (6) have a shape of disks 3–6 μm in diameter.

As it follows from data of the quantitative microstructural analysis carried out according to the authors' method [69], the pore space of a clay sediment having the honeycomb microstructure is represented by four categories of

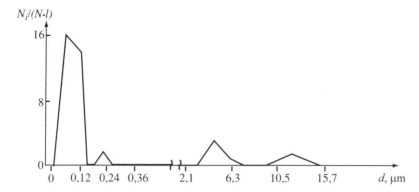

Figure 1.13. The curves of pore distribution according to the equivalent diameters for the honeycomb microstructure.

pores with average equivalent diameters of 0.06, 0.22, 5.3, and 11.6 μm (Fig. 1.13). The number of pores in each size range may be estimated using the probability density ρ:

$$\rho = N_i/(N \times l) \tag{1.23}$$

where N_i is the number of pores in the given range; N – total number of pores; l – the range value. Here and below, the probability density values will be taken from the curves of pore distribution according to the equivalent diameters.

In the honeycomb microstructure, small isometric micropores 8 (see Fig. 1.12, *a*) with the average equivalent diameter of 5.3 μm are prevalent among the intermicroaggregate pores (probability density 3.2, Fig. 1.13). Large isometric pores 9 (see Fig. 1.12, *a*) with the average equivalent diameter of 11.6 μm are much lesser in number (probability density 1; Fig. 1.13).

Interparticle ultramicropores 3 (see Fig. 1.12, *b*) are the most numerous (probability density 16; Fig. 1.13). The average equivalent diameter of these pores is 0.06 μm. The interparticle ultramicropores are anisometric, for they are formed as a result of a loose contact between the basal surfaces of platy particles.

Interultramicroaggregate fine micropores 4 (see Fig. 1.12, *b*) are less numerous (probability density 2.7, Fig. 1.13). They have the average equivalent diameter of 0.22 μm. These pores are also anisometric.

Despite the predominance of the interparticle and interultramicroaggregate micropores, the summary value of their area is insignificant. The intermicroaggregate micropores (with equivalent diameters of 4–12 μm), although less in number, compose the bulk of the pore space in clay sediments with honeycomb microstructure. The quantitative analysis shows that the summary area of these pores may be as high as 97% of the total measured porosity (the total porosity of the sediment is ~57%).

The results of the orientation analysis show the predominantly isotropic character and the absence of any orientation of structural elements in the honeycomb microstructure, the evidence of which is provided by the circular character of the orientation rose (Fig. 1.12, c) and small value of the anisotropy coefficient ($A_g \sim 4.1\%$) [23].

The clay sediments with honeycomb microstructure have high porosity (up to 60–90%). Their natural water content exceeds the water content at the liquid limit and attains 55–200%.

As has been already said above, the interaction of particles inside the microaggregates and ultramicroaggregates in such sediments is realized through the close coagulation contacts (aggregation in the close potential minimum), while the distant coagulation contacts are dominant between the microaggregates.

2

Lithogenesis of Clay Sediments

2.1. GENERAL INFORMATION

Lithogenesis is commonly understood as all the changes in sediments and in the forming sedimentary rock up to the beginning of its metamorphism. In the course of lithogenesis, sediments pass through a number of alteration stages to become cemented sedimentary rocks.

There are differences in the understanding of the lithogenesis process and its subdivision into separate stages. Many specialists-lithologists use the concept of diagenesis in describing all the alterations in sediments and sedimentary rocks. This term was proposed for the first time in 1883–1888 by German geologists K.W. Gümbel and I. Walter. Subsequently, it became widespread not only in Germany, but also in England, France, the USA [2, 72, 73].

Diagenesis is translated as «regeneration» or «transformation», which allows its use in order to characterize the whole complex of processes forming the basis of the sedimentary rock formation. Later, however, mainly in the works of Russian geologists, this term began to be used to characterize only the initial lithogenesis stage, and namely, the stage at which loose unconsolidated sediment was transformed into consolidated sedimentary rock. Such an approach has been worked out in the works of A.E. Fersman [74]. A.G. Kossovskaya et al. [75], N.B. Vassoevich [76], N.M. Strakhov [77, 78], N.V. Logvinenko and L.V. Orlova [79], and many others.

Further alterations in the rock, which develop after the diagenesis is over, have been termed *catagenesis*, as suggested by A.E. Fersman [74]. This concept became widely used by N.B. Vassoevich [76, 80], N.V. Logvinenko [81], V.D. Shutov [82], W.S. Fife et al. [83]. Other scientists began to use the term *epigenesis* in describing the postdiagenetic transformations in rocks [84–86]. As the term «epigenesis» is used in the ore geology in another sense, the first of the terms has become preferably used.

Later, at the boundary between catagenesis and metamorphism, one more stage has been differentiated and given the name of *metagenesis* [75]. N.M. Strakhov [87] used this concept in a wider sense, including in it all the alterations of rocks from diagenesis through metamorphism, and N.B. Vassoevich [76] proposed to use the term «metagenesis» instead of «metamorphism». Later on, however, the majority of researchers stuck to the original understanding of this term [79, 88].

The above stages are considered in the formation of sedimentary rocks of various composition: clay, sand, carbonate, siliceous. In petroleum geology, the subaqueous lithogenesis of clay sediments is of primary interest and will be the subject of the following description.

2.2.   THE STAGES OF LITHOGENESIS

2.2.1.   DIAGENESIS

Diagenesis, as it is presently understood by the majority of lithologists, is the transformation of a loose sediment into consolidated deposits (plastic clays) under the influence of mainly biochemical and physicochemical processes.

All diagenetic transformations proceed under the pressure-temperature conditions similar to those of the sediment formation. The temperature change in the diagenesis zone does not exceed several degrees. The water column pressure may vary depending on the conditions of sediment accumulation from zero to 3 MPa, while the geostatic pressure of the sediment proper does not exceed 3–4 MPa.

Depending on the depositional environments, the forming sediments may differ in their mineral composition and grain size distribution, content of organic matter, composition and salinity of pore waters, pH and Eh of the medium. Therefore, the process of lithogenesis, although showing a general trend, has its peculiarities in every sedimentary basin.

Lithogenetic transformations begin practically immediately after the sediment formation. An important role at the initial stage is played by microbiological processes. In the superficial sedimentary beds containing organic matter, bacterial activity results in creating the alkaline oxidizing medium, in which the processes leading to the formation of iron and manganese oxides and hydroxides (iron-manganese concretions) intensely develop; if pyroclastic material is present, iron smectites and zeolites may form.

As the sediment becomes buried, the biochemical processes develop under the alkaline reducing conditions and result in the reduction of trivalent iron and manganese oxides into the bivalent compounds, in the reduction of pore water sulfates with the formation of hydrogen sulfide, iron and sometimes manganese sulfides, and in the dissolution of carbonates. Conditions are created favoring the formation of authigenic minerals such as pyrite and hydrotroilite. The alteration of clastic feldspars, micas, clay minerals results in the formation of hydromicas 1M. Transformation of volcanic material may serve as a source for the formation of montmorillonite and mixed-layer minerals (Fig. 2.1).

The bulk of clay minerals in the formed sediments are, however, allothigenic. A characteristic feature is the high content of dioctahedral-type clay minerals: kaolinite, halloysite, montmorillonite, hydromica $2M_1$, montmorillonite-mica mixed-layer varieties containing more than 10% of labile layers.

Physicochemical processes are of no less importance in the transformation of sediments at the early diagenesis substage. The source of energy for them is provided by the physicochemical and biological transformations releasing the required quantities of energy. The basis of the physicochemical compaction is

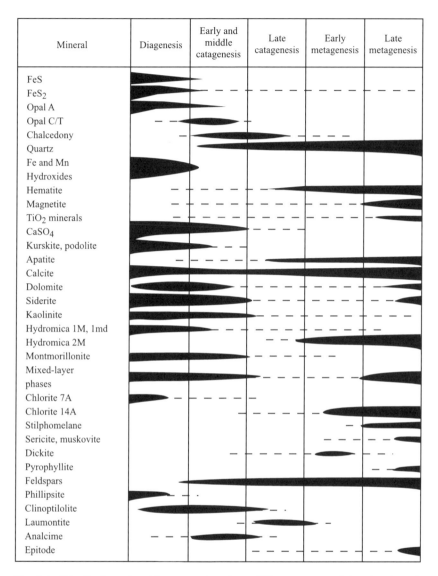

Fig. 2.1. Distribution of authigenous minerals at different stages of the clay deposit formation and alteration [79].

formed by a series of interconnected processes, the most important of which are the following: aging of alumo-silica gels and aggradation of clay minerals, changes in the composition of pore solution and exchange complex, transformation of organic matter, syneresis.

Investigation of young sediments and laboratory experiments on the synthesis of clay minerals [73] show that alumo-silica gels having high wetting ability and porosity are formed in the process of sediment accumulation.

With time, the gels become more ordered, with shapeless aggregates of fine particles being converted into platy scales having two-dimension arrangement and more compact structure.

Along with the structural ordering of gels, the process of clay mineral aggradation develops, which starts immediately after the sediment accumulation due to the elevated contents of magnesium and potassium ions in ooze waters. The basis of this process is the «healing» of crystal lattices of clay minerals that have undergone considerable degradation during weathering and transportation, through the fixation of magnesium, iron, and potassium cations inside the layers and in the interlayer space [73, 89]. Such a transformation of minerals, proceeding intensely also at later stages of diagenesis and in catagenesis, increases the degree of clay mineral crystallization, leads to the change in the ratio of their swelling and non-swelling varieties.

Investigation of recent sediments shows that the composition and concentration of ooze waters may differ considerably from those of basin waters. The scale of these alterations is different and depends on a number of factors: grain size distribution and mineral composition of the sediment, content of organic matter, intensity of biochemical processes, content of carbonate material, etc. One of the most characteristic trends in the diagenetic transformations of ooze waters is an increase in their Mg^{2+}, Na^+, and K^+ contents, resulting in considerable alterations occurring also in the exchangeable complex of the sediments. A decrease in the wetting ability of clay particles and in the degree of diffusion of their electrical double layer are connected with these processes. Therefore, the observed diagenetic transformation of the liquid component of sediment may in the end influence its compaction and water content [90].

An important role in the sediment diagenesis is played by organic matter. When organic remains are intensely supplied to the sediment, they may be buried in the form of turf and sapropel. However, the majority of dispersed vegetable remains occurring in sediments become quickly decomposed. Decomposition of organic matter begins with the autolysis of dead cells manifested in the hydrolysis of organic matter molecules and their disintegration resulting in the formation of acids and alcohols. Further decomposition of cellular tissue proceeds under the influence of fungi and bacteria. This results in the formation of CO_2 and CH_4 under aerobic conditions, while under anaerobic conditions CO_2, CH_4, H_2, and fatty acids form. Lignin is transformed into humic acids, albumen are sources for the formation of CO_2, NH_3, H_2S, and amino acids. Organic matter loses its cell structure and turns into a watered colloid.

Humic acids and their soluble humates produced from the organic matter decomposition are actively adsorbed on the surface of clay minerals where they form organic-mineral complexes. This modifies the surface of minerals (makes it hydrophilic), increases the stabilization of particles and weakens their structural cohesion with one another. As a result, favorable conditions are created for the mutual displacement of particles and spontaneous compaction of the sediment.

In the process of diagenesis, the contents of humates and humic acids increase in a regular way, while the quantities of fulvic acids and bitumen significantly decrease. The fulvic acids and bitumen, which are not connected with the surface of clay minerals, are probably squeezed out of the sediment together with pore solution, and so migration of hydrocarbons begins already at the stage of diagenesis.

In the process of diagenesis and at the following lithogenesis stages, the organic matter remaining in the sediment becomes more and more tightly fastened on the surface of clay minerals. At the same time, the high-molecular compounds of humic acids become condensed, lose their wetting ability and turn into carbonized organic matter.

All the above processes cause the «aging» of clay sediments, which develops intensely at the initial stages of their transformation and is usually seen from a decrease in the colloidal activity of the fine grained fraction downward the section. The decrease in the wetting ability results, in its turn, in the spontaneous compaction of the sediment and its dehydration, in the increase of contact interactions and strength of the system.

The sediment condition is strongly influenced by the process of syneresis – spontaneous dehydration and strengthening of the structure. The cause of syneresis is the absence of true thermodynamic equilibrium in the mutual fixation of particles and aggregates at great distances. This determines the relaxation nature of further processes, which is expressed in the syneresis. The basis of syneresis is a gradual thinning of the hydrate film at the contact of particles caused by the action of molecular forces. This effect has been studied for the first time by B.V. Deryagin and A.D. Malkina for the case of two quarts threads interacting in a liquid medium [41]. They have registered a gradual decrease in the distance between the threads with time and an increase in the molecular attraction forces.

In spite of the processes of aging and syneresis, general concentration of mineral particles in a unit volume of the sediment at the early stage of its diagenesis remains low as before. Sufficiently thick hydrate films are preserved at the contacts of particles to bring about a stabilizing effect opposing the molecular attraction forces. So the system as a whole continues to be of the coagulation type with all the properties inherent to it: high porosity and water content (higher, as a rule, than that of the liquid limit), well expressed thixotropy, low strength, etc.

With the progress of diagenesis, the role of biochemical and physicochemical factors gradually lessens, activity of bacteria ceases, water exchange between the pore solution and the above-bottom waters becomes impeded, the role of gravitational compaction gradually grows. The intensity of gravitational (geostatic) compaction depends on the character of the sediment, degree of its homogeneity, rate of pore water expulsion, and time of compaction. The highest rate of diagenetic compaction is shown by the coarser in composition and less homogeneous in structure clay sediments formed in the shallow-water zones of intracontinental basins and oceanic shelf. Development of heterocoagulation processes in these zones results in the formation of sediments having heterogeneous structure and high filtration characteristics. Interbeds of sand and shell deposits often present in such sediments also contribute to the pore water drainage and accelerated compaction even at minimal gravitational loading.

In deeper-water parts of seas and oceans, conditions for the gravitational compaction of sediments in the course of diagenesis change significantly. Stabilization of clay particles by organic compounds, in combination with the high homogeneity and fine grain size of sediments, favor the formation of a homogeneous honeycomb structure that gives away water with difficulty. The coefficient of pore water filtration in such sediments does not exceed 0.01 cm per year. Besides, with an increase in basin depth, the pore pressure grows at the expense of the water column weight, which also hinders the accelerated compaction of deep-water sediments [79].

The above is confirmed by observations of the time required for the development of diagenetic processes in sediments accumulated in different depositional environments. Thus, in the littoral and shallow-water shelf zone, sediments that have undergone diagenetic transformations are encountered already among recent sediments (the time of transformations is several thousand years), on the continental slope they are encountered among the sediments of Paleogene and Neogene age (several ten million years) and at the abyssal depths – in the Cretaceous and Jurassic (about hundred million years) [79].

The newly formed sediments have, as a rule, a high-porosity structure, whose interaggregate pores (cells) are filled with free pore solution. Therefore, free (gravity) water is prevalent in the sediment at the early substage of diagenesis. Owing to its free connection with the near-bottom solution, the concentration of salts in the pore water remains similar to that of the external solution. As has been already mentioned above, some difference between the external and the pore solution is observed in the composition of ions and cations.

In the process of physicochemical and especially gravitational compaction, free water is gradually squeezed out of the pore space, being preserved only in large isolated pores, which can remain in the sediment structure because of the low value of geostatic loading.

By the end of the diagenesis stage, physically bound (osmotic and adsorptionally bound) water becomes the predominant type of water in the sediment. At the same time, the role of the osmotic transfer of water molecules and salts dissolved in water increases. Cations and anions transported towards the sediment from the bottom waters are spent for the authigenous neocrystallization inside the sediment, while salts that have undergone dissolution leave the sediment.

As the sediment is buried to greater depths and its density increases, changes in its microstructure occur. The main trend is the drawing together and enlargement of microaggregates, closing of large and reduction in size of small micropores. Small intermicroaggregate micropores (1–10 μm) become prevalent. All this causes the transformation of clay sediment microstructure: the honeycomb microstructure predominant in young sediments (see Fig. 1.12) is transformed at first into the transitional honeycomb-matrix and then into the matrix microstructure. Porosity in the course of diagenesis diminishes from 60–75% at the early substage to 35–45% at the late substage.

The *matrix microstructure* that forms in diagenesis has been described by us earlier for clays of various genesis [21, 69, 91]. The main characteristic feature of this microstructure is the presence of a continuous clay mass (matrix), which contains randomly located silt and sand grains not contacting between themselves.

As an example of the matrix microstructure, Figure 2.2, *a* shows the photomontage of SEM images of a marine clay sample (*m* Pg$_2$) recovered in the Tyumen' region near the village of Pershino. As can be seen from the figure, the principal solid structural elements of the matrix microstructure are

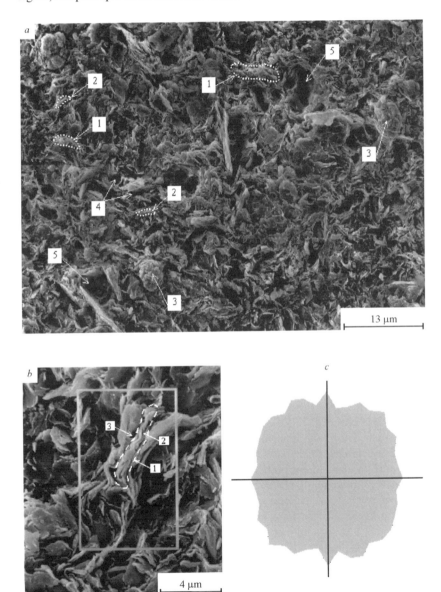

Fig. 2.2. General view (*a*) of the matrix microstructure of a marine clay (Tyumen' region, Pershino village), its fragment (*b*), and the rose of orientation of structural elements (*c*).

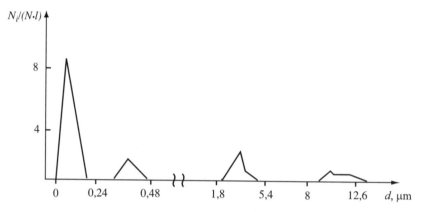

Fig. 2.3. A curve of pore distribution according to equivalent diameters for the matrix microstructure.

microaggregates (1) having sizes of 4–15 μm and smaller elongated microaggregates (2) whose thickness may attain 2 μm and length – 6 μm. The microaggregates have a complicated structure (Fig. 2.2, *b*) and consist of clay particle ultramicroaggregates (1) contacting according to the face-to-face and face-to-edge type. The ultramicroaggregates are anisometric in form and composed of clay particles contacting by their basal planes. The thickness of ultramicroaggregates ranges from 0.3 to 0.6 μm and their length is 2–3 μm.

Silt grains (3) are usually isometric and their surface is covered by clay «coats» (Fig. 2.2, *a*). The grain size ranges from 5 to 15 μm.

A quantitative analysis of the microstructure accomplished using SEM images after the authors' technique [69] has shown that the pore space of clays with the matrix microstructure is represented by four categories of pores with average equivalent diameters (d) of 0.06, 0.42, 3.3, and 12.3 μm (Fig. 2.3). The pore space is largely formed by intermicroaggregate pores making up as much as 93.8% of the total porosity (total porosity of the sample is 39%). Among the intermicroaggregate pores, small micropores (4) are prevalent (probability density 2.2) (see Fig. 2.2, *a*). The average equivalent diameter of such pores is ~3.3 μm. They have an isometric or anisometric form.

Large isometric micropores (5) (Fig. 2.2, *a*) with $d \sim 10.9$ μm are encountered much more seldom (probability density 0.3).

The intramicroaggregate porosity represented by interparticle (2) and interultramicroaggregate (3) pores (Fig. 2.2, *b*) is of subordinate importance in the matrix microstructure. The sum of these pores amounts to 6.2% of the total porosity.

The interparticle anisometric ultramicropores with $d \sim 0.06$ μm are the most numerous (probability density 9.5) in the matrix microstructure.

The interultramicroaggregate fine anisometric micropores (3) (see Fig. 2.2, *b*) having $d \sim 0.42$ μm are less widespread (probability density 1.85).

The results of investigations point out the isotropic character of orientation in the matrix structure and the absence of the preferred orientation of structural elements. This is evidenced by the circular form of the orientation rose (see Fig. 2.2, c) and by the small coefficient of anisotropy ($A_g = 5.9\%$) [23].

In comparison with the quantitative analysis results obtained for the honeycomb microstructure of young clay sediments (see Fig. 1.13), the matrix microstructure shows a noticeable decrease in the content of large micropores (more than 10 μm). At the same time, the number of small micropores (1–10 μm) increases, though their average equivalent diameter becomes a little less than in the honeycomb microstructure. The size of fine interultramicroaggregate micropores (0.1–1 μm) in the matrix microstructure shows a little increase, obviously at the expense of destroyed large and small micropores that have passed to this category. The interparticle ultramicropores (<0.1 μm) undergo the least changes. Both in the honeycomb and matrix microstructures, this is the most numerous category of pores, whose equivalent diameter remains equal to 0.06 μm.

The physicochemical and gravitational compaction of clay sediments results in the growth of effective stresses at the contacts and in the overcoming of the energy barrier with the passage of particles from the distant potential minimum to the close one. A calculation of the minimum effective stress required for the conversion of the distant coagulation contact into the close one may be done based on the assumption that the energy barrier may be overcome if the effective stress from the overlying sediments exceeds its value. The stress required to overcome the energy barrier may be found from Equation (1.22) with the use of experimental data obtained by J. Israelachvili and G. Adams [28] according to which the unit energy of repulsion forces at the energy barrier is equal to $F(h)/2 = 1$ mJ/m^2. Knowing the value of the critical effective stress at a contact (P') and the number of contacts in 1 cm^2 of the sediment, one can find the value of the external pressure required for overcoming the energy barrier and transforming the distant coagulation contacts into the close ones.

The calculations performed show that the minimum pressure value is 10 kPa. Consequently, the conversion of the distant coagulation contacts into the close ones begins already when the sediment thickness is about 1 m. It may be suggested on these grounds that structural cohesion in the early diagenesis subzone is determined by the distant and close coagulation contacts. With the passage to the late diagenesis subzone, the probability that the distant coagulation contacts will be preserved is very low. It is obvious that the close coagulation contacts become predominant in this subzone.

When the close coagulation contacts begin to dominate, the strength of the sediments increases, their condition and the character of behavior in deformation change: sediment deformation acquires a pronounced viscous-plastic character, thixotropic properties gradually decrease.

2.2.2.   CATAGENESIS

Catagenesis is understood as the totality of processes causing the transformation of consolidated sediments into indurated sedimentary rocks under elevated pressures and temperatures as the rocks subside to deeper horizons of the Earth's sedimentary cover. The zone of catagenesis has a considerable thickness extending as deeply as 10 km. The geostatic pressure in this zone attains 120–200 MPa, temperature is 150–200 °C [79, 92, 93].

The principal factors of rock transformation at the catagenesis stage are the high pressure and temperature. Therefore, the processes of catagenesis are physicochemical and physical-mechanical in their nature, the role of biochemical factors at this stage of lithogenesis becomes significantly reduced.

Catagenesis of clay sediments develops in the sedimentary cover of platforms, in platform syneclises, in the foredeeps and upper structural stage of geosynclines, as well as in sedimentary rocks on the oceanic crust. Development of catagenesis has its specific features in different depositional environments. This is determined by the great variety of pressure-temperature and physicochemical conditions. For example, the geothermal gradient on platforms averages about 33 m/degree, in geosynclines it is 15–25 m/degree, and in the sedimentary cover of oceans it is still less (14–18 m/degree). Thus, depending on the geologic structure, the same temperature may be encountered at different depths, which affects the development of catagenetic processes in these structures.

The following processes occur during catagenesis: rock compaction, alteration of the composition and content of pore waters, mineral formation, crystallization of colloids and amorphous substances, recrystallization of minerals, transformation of structure and strengthening of structural bonds.

Catagenesis develops under the conditions of continuously growing geostatic pressure caused by the loading of overlying beds. In geosynclines, as well as in platform areas where dome-like structures and flexures are developed, there is also the stress pressure superimposed on the geostatic pressure, under whose influence cleavage may form. This explains the fact that rocks in the geosynclinal areas show, as a rule, a little greater compaction than on the platforms. However, in the most part of the catagenesis zone, only the geostatic pressure acts. In spite of its huge value amounting to 200 MPa, this pressure is not sufficient for crushing the clastic grains and, all the more, the fine grained minerals of clay rocks [94–95]. So the compaction proceeds mainly at the expense of the mutual displacement of structural elements and their closer packing. Catagenetic transformations of rocks develop at the slowest rate under the conditions of great oceanic depths. Deep-water drilling has made it possible to recover from oceanic bottom the rock samples occurring only at the early and middle substages of catagenesis. This is explained by the insignificant growth of geostatic pressure in deep-water oceanic rocks because of the low rate of sediment accumulation.

In the course of compaction, the squeezing-out of pore water continues. At the early substage of catagenesis, the free water contained in large pores and the osmotically bound water are squeezed out. As, together with water, ions of the diffuse layer of clay particles are also squeezed out, this results in an increase in the squeezed-out pore solution concentration.

The squeezed-out pore solution moves vertically along the section to the silt and sand horizons. This phenomenon is common in the shallow-water clay sediments usually containing the draining interlayers. In homogeneous clay sequences, the squeezed-out water may move horizontally towards the major tectonic faults or sand-enriched facies. This is favored by the filtrational anisotropy of clay sequences and their higher along-bed permeability due to the higher degree of orientation of clay structural elements normally to the acting load.

In the absence of draining interlayers, the removal of squeezed-out water from clay rocks is impeded, which causes the rise of anomalously high formation pressure (overpressure). The complete break of hydraulic connection between the pore water and the external hydrosphere results in the increase of stresses inside the rock due to the geostatic pressure.

Temperature exerts a great influence on the compaction of rocks and chemical–mineral transformations occurring in them. The role of temperature in the development of catagenesis processes is manifested first of all through the change in the state and activity of water contained in the rock. A growth of temperature increases water activity: its dissociation degree, mobility, and dissolving power grow. Dissolution of amorphous silica increases 5–6 times with the growth of water temperature from 0 to 80–90 °C. Quartz, practically insoluble under usual conditions, at a temperature of about 200 °C acquires solubility equal to several mg/l. A decrease in water viscosity increases significantly the permeability of rocks to water and accelerates their dehydration during compaction: as the temperature grows to 300 °C, the permeability of clay rocks to water increases almost 10 times [96].

A significant effect on the development of rock compaction is produced by the transformation of adsorptionally bound water into free water. Experimental investigations by A.A. Blokh [97], B.V. Deryagin and N.V. Churaev [12], Y.I. Tarasevich and F.D. Ovcharenko [98] have established that at a temperature of 65–70 °C, the increase in the translation movement of water molecules results in the weakening of their connection with the mineral surface and, as a consequence, the physical condition of water changes: the adsorbed water turns into the free (gravity) water. As the adsorption centers retaining the adsorptionally bound water are energetically heterogeneous, one may expect that the transformation of tightly bound water into free water will be observed at temperatures of 65–110 °C and higher. It should be noted that the external pressure tends to prevent the growth of the translation movement of water molecules. So the influence of temperature and pressure on the condition of adsorptionally bound water in the catagenesis zone are oppositely directed.

The transformation of adsorptionally bound water into free water may also occur in the course of mineral alterations. In particular, the process of montmorillonite hydromicatization that develops in catagenesis is accompanied by the liberation of adsorbed water molecules hydrating the interlayer space of montmorillonite.

The passage of adsorptionally bound water into free one results in the acceleration of rock dehydration and compaction, as well as in the appearance in the pore solution of free water with enhanced dissolving power caused by the protonization effect. The latter is related to the partial dissociation of the formed free water at elevated temperatures [8, 98]. The presence of free water with enhanced dissolving power activates the process of dissolution of salts and some even poorly soluble minerals. This is confirmed by the decrease in the salt contents in clay rocks occurring below the isotherm of 65–70 °C.

Along with the compaction, the processes of mineral formation and colloid aggregation continue in catagenesis. Their characteristic features are the isolation in pores of neocrystallized forms such as grains, grain aggregates, and crystals of a regular form and the recrystallization of the earlier isolated fine-crystalline compounds. The most significant alterations of clay mineral composition are those occurring chiefly at the middle and final substages of catagenesis. They are manifested in the hydromicatization of montmorillonite with the formation of dioctahedral micas. The process passes through the formation of a transitional phase – mixed-layer minerals of the hydromica-montmorillonite and hydromica-montmorillonite-chlorite series (see Fig. 2.1).

The transformation of montmorillonite into hydromica in the course of catagenesis has been for the first time noted by C. Weaver [99]. Then J. Burst [100] and M. Power [101] have studied the main regularities of this process. Using the Tertiary clay section in Texas as an example, they have established that the transformation of montmorillonite into hydromica through the mixed-layer phase occurs in the depth interval from 2250 m to 3750 m and that at depths greater than 4200 m the swelling minerals are completely absent in the clay rocks. Later, the post-sedimentary transformation of clay minerals in lithogenesis has been studied by A.G. Kossovskaya [102], V.D. Shutov [82], I.D. Zkhus and V.V. Bakhtin [103]. The data on the changes in the clay composition under natural conditions have been confirmed by laboratory experiments [104]. It has been established that the process of mineral transformation has its peculiarities in each basin. For example, within the Jurassic and Cretaceous sedimentary section of Central Dagestan, hydromica was found to appear at a depth of 3700 m [105]. On the Stavropol uplift, the boundary at which hydromicatization of montmorillonite begins was established in Cretaceous sediments at a depth of 2.3–3.3 km. At the same time, in the middle Pliocene-Quaternary sediments of the South-Caspian basin, the transformation of montmorillonite into hydromica was not observed at depths shallower than 5–6 km [106].

Generalization of data on the change in the clay mineral composition of seals at the stage of catagenesis has been accomplished by [107], with reference to the clays of West Siberia (Fig. 2.4). As one can see from these data, there is a noticeable increase with depth in the contents of mixed-layer minerals and hydromica, with a concurrent decrease in the montmorillonite content. The latter begins to play a subsidiary role and practically disappears already at depths of about 1500 m. At the same time, at these and greater depths, the amount of mixed-layer minerals increases, with the swelling component content in them gradually decreasing. At depths over 3000 m, hydromica, mixed-layer varieties (with the content of swelling components not exceeding 30%), and chlorites become the prevailing minerals in the seals.

No less interesting results have been obtained by I.D. Zkhus and V.V. Bakhtin [103] in studying the composition of clay minerals from marine and continental deposits that had undergone various lithogenetic transformations (Fig. 2.5). These authors have established the typical complexes of clay minerals for humid and arid zones at different substages of diagenesis and catagenesis; in the late catagenesis, these complexes acquire similar features and lose their original

Depth of rock occurrence, m
Mineral composition of clays
Montmorillonite
Mixed-layer mineral
Hydromica
Chlorite
Kaolinite
1 2 3
Content of swelling layers in mixed-layer minerals,%
20 40 60 80
500
1000
1500
2000
2500
3000
3500

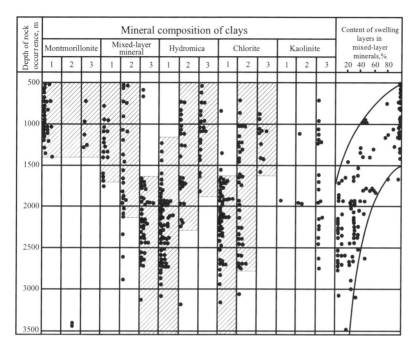

Fig. 2.4. Change in the composition of clay minerals with depth in the Mesozoic clay seals of West Siberia [107]. Relative contents of minerals in clays: 1 – prevailing, 2 – subordinate, 3 – insignificant. Hatched: zones of the prevailing distribution of clays with some or other relative content of clay minerals.

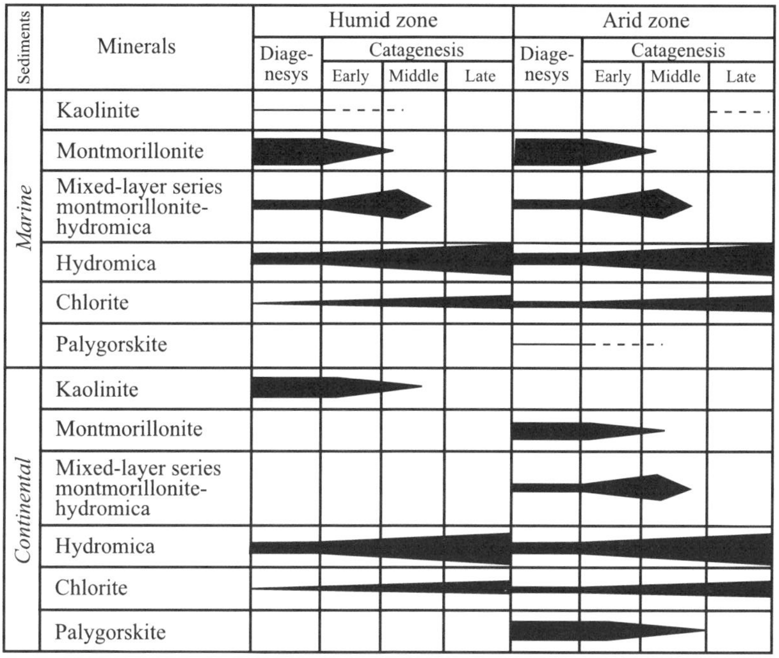

Sediments
Minerals
Humid zone
Arid zone
Diage-nesys
Catagenesis
Early
Middle
Late
Marine
Kaolinite
Montmorillonite
Mixed-layer series montmorillonite-hydromica
Hydromica
Chlorite
Palygorskite
Continental
Kaolinite
Montmorillonite
Mixed-layer series montmorillonite-hydromica
Hydromica
Chlorite
Palygorskite

Fig. 2.5. Change in the clay mineral composition in sediments of the humid and of the arid zone at different stages of lithogenesis [103].

individual peculiarities. Thus in the late catagenesis, the associations of clay minerals, independently of their original compositions, become predominantly two-component, consisting of hydromica and chlorite.

N.B. Vassoevich [108] has studied the interconnection between the alterations of clay minerals in the course of lithogenesis and the transformation of organic matter into coals. It has been established that the zone of montmorillonite existence extends to the depth, at which «gas» coals are formed (2500–3300 m), and the maximal development of mixed-layer minerals corresponds to the pressure–temperature conditions, at which «long-flame» and «gas» coals are formed (1650–3300 m). At greater depths corresponding to the formation of «fat» coals (3300–4000 m), a gradual decrease in the quantities of mixed-layer minerals with a high content of swelling components is observed.

In a number of cases, vitrinite reflectance index (R^a) may serve as a quantitative criterion of the degree of clay rock alteration at the stage of catagenesis. The value of this index grows regularly with depth depending on the degree of coaly matter transformation (Fig. 2.6) [109]. However, it should be taken into account in using this method that, the general trend of R^a change with depth remaining valid, its numerical values may differ for rocks of different ages. Besides, the coal-petrographic investigations become often impossible because of the absence of coal inclusions in cores.

At the stage of catagenesis, further compaction of clay rock and transformation of its microstructure occur as a result of increasing pressure and

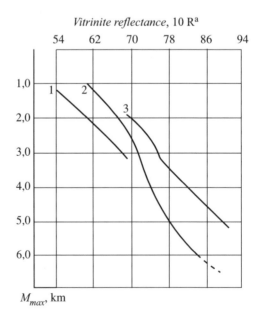

Fig. 2.6. Change in the vitrinite reflectance depending on the depth of clay deposit occurrence [109]. Age of sediments: 1 – Pliocene, 2 – Miocene, 3 – Paleogene to Upper Cretaceous.

temperature. These transformations are expressed in the compaction and enlargement of ultramicroaggregates and aggregates, their gradual transformation at first into large elongated microblocks and then into polycrystalline overgrowths up to 10–12 μm thick. The degree of orientation of solid structural elements increases and the anisotropy of the physical and mechanical properties of clay rocks grows. Owing to the intense pressure and the «intergrowth» of clay particles inside the microblocks and polycrystalline overgrowths, the ultramicropores disappear and the number of interultramicroaggregate micropores decreases. At the same time, the number of anisometric and fissure-like micropores separating the microblocks and overgrowths increases.

Under the influence of the occurring alterations, the earlier-formed microstructures become gradually transformed and new structures are formed. Turbulent and laminar microstructures are most common at the stage of catagenesis. The first of them forms at the initial substage and the second – at later substages of catagenesis.

The *turbulent microstructure* is presented in Figure 2.7, *a* showing a photomontage of SEM images of a Cambrian (ε_1) marine clay sample. The most characteristic feature of such a microstructure is the presence of leaf-like curved aggregates of clay particles oriented along the bedding, which flow round silt grains like a turbulent flow.

The principal solid structural elements (Fig. 2.7, *a*) of the turbulent microstructure are anisometric microaggregates oriented along the bedding. The microaggregates (1) often curve to repeat the form of silt grains which they flow around. The length of such microaggregates may reach 20 μm and their thickness is 2–3 μm. The microaggregates have a complicated structure and consist of platy and leaf-like ultramicroaggregates (1) (Fig. 2.7, *b*), which contact by their basal planes and according to the face-to-edge at a low angle type. The length of the ultramicroaggregates may attain 3 μm and their thickness varies from 0.2 to 0.6 μm. The ultramicroaggregates consist of clay mineral particles contacting by their basal planes.

Besides the microaggregates and aggregates, the turbulent microstructure may contain individual inclusions of silt grains (2) having a predominantly isometric form (Fig. 2.7, *a*), whose size may vary in rather a wide range – from 5 to 20 μm. The surface of grains is usually covered with clay «coats».

A quantitative analysis of the microstructure carried out using SEM images shows that pore space in clay rocks having turbulent microstructures is represented by four categories of pores with average equivalent diameters (d) of 0.06, 0.26, 3.3, and 10.2 μm (Fig. 2.8). The most part of the pore space is composed of small intermicroaggregate (3) (Fig. 2.7, *a*) and fine interultramicroaggregate (3) (Fig. 2.7, *b*) micropores that make up 93.7% of the measured total porosity (the total porosity value for this rock is 28.7%).

Large intermicroaggregate micropores (4) (Fig. 2.7, *a*) in clay rocks with turbulent microstructure are encountered very seldom, mainly at the places of junction between clay microaggregates and silt particles. They have an isometric form and $d \sim 10.2$ μm.

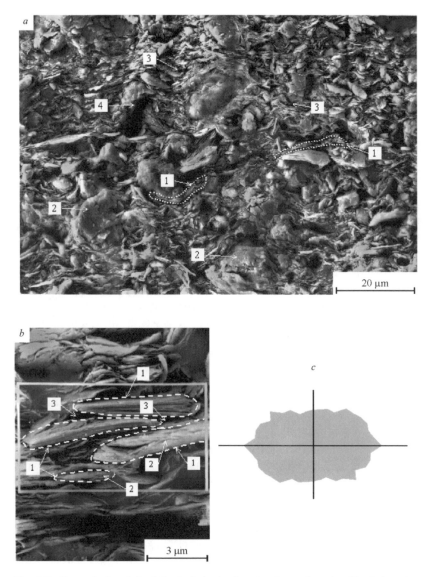

Fig. 2.7. General view (*a*) of the turbulent microstructure of marine Cambrian clay (St. Petersburg area), its fragment (*b*), and orientation of structural elements (*c*).

The small intermicroaggregate micropores form the second (after ultramicropores) in multiplicity category of pores in the given microstructure (probability density 3.8). These are, as a rule, anisometric pores with $d \sim 3.3$ μm.

The interultramicroaggregate fine anisometric micropores (3) (Fig. 2.7, *b*) with the average equivalent diameter of 0.26 μm are a little less distributed (probability density 2.4).

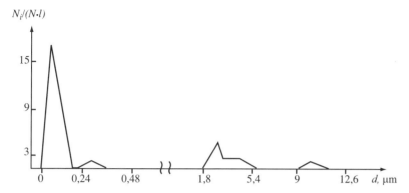

Fig. 2.8. A curve of pore distribution according to equivalent diameters for the turbulent microstructure.

The interparticle ultramicropores (2) of a fissure-like form (Fig. 2.7, *b*) with $d \sim 0.06 \, \mu$m are the most numerous (probability density 15.9). It is necessary to note that the equivalent diameter value for this category of fissure-like pores does not give a complete idea of their size, so it is more correct to use the area values or to indicate the maximal width of opening of such pores.

An analysis of the turbulent microstructure has shown that it ranks with the category of high-oriented microstructures with the along-bedding preferred orientation of structural elements. This is also confirmed by the ellipse-shaped character of the orientation rose and by the high value of the coefficient of anisotropy ($A_g = 28.7\%$) (Fig. 2.7, *c*).

The difference of the turbulent microstructure from the matrix microstructure formed in diagenesis lies in the growing role of small and fine micropores in the formation of pore space in clays. Large pores practically disappear. The number of ultramicropores remains relatively high, as before. The orientation of structural elements becomes considerably higher causing the physical, filtrational and mechanical anisotropy of clay rocks.

A characteristic feature of the *laminar microstructure* is the high degree of structural element orientation in the bedding plane. This is an example of a typical homogeneous and anisotropic microstructure. Figure 2.9 show the photomontage of SEM images of argillite (K_1).

The principal solid structural elements (Fig. 2.9, *a*) of the laminar microstructure are anisometric microaggregates (1) oriented along the bedding. The length of the microaggregates varies from 7 to 15 μm and their thickness is about 1–2 μm. The microaggregates have a complex structure and consist of leaf-like ultramicroaggregates (1) (Fig. 2.9, *b*) contacting by their basal surfaces. The ultramicroaggregates are 2–6 μm long and 0.2–0.5 μm thick. They are formed by clay particles contacting according to the face-to-face type.

Besides the microaggregates and ultramicroaggregates, polycristalline microaggregate overgrowths of anisometric form, also elongated along the bedding, are present in the laminar microstructure. Their maximal thickness may attain 3–4 μm.

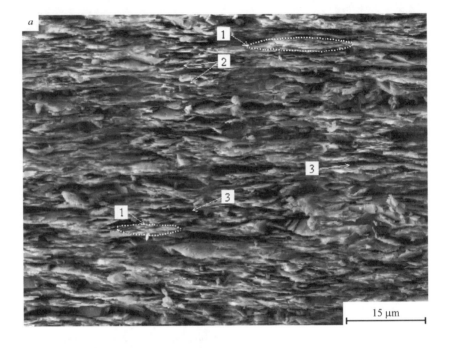

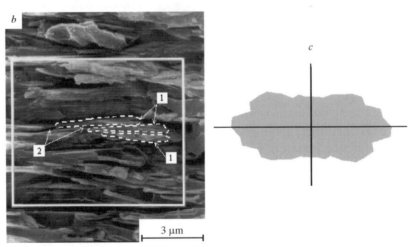

Fig. 2.9. General view (*a*) of the laminar microstructure of argillite (Sochi area), its fragment (*b*), and orientation of structural elements (*c*).

A quantitative analysis of SEM images of the laminar microstructure shows that the pore space of clay rocks with such a microstructure is represented by three categories of pores (Fig. 2.10) with average equivalent diameters of 0.11, 1.5, and 3.4 μm.

The main part of the pore space is formed of small intermicroaggregate micropores with equivalent diameters of 1.5 and 3.4 μm. They make up 71% of the

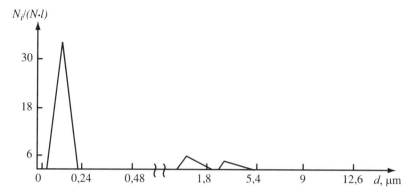

Fig. 2.10. A curve of pore distribution according to equivalent diameters for the laminar microstructure.

measured total porosity (the total porosity value for the given rock is 9.3%). These micropores (2) and (3) (Fig. 2.9, *a*) have fissure-like and anisometric forms.

Fine interultramicroaggregate micropores (2) (Fig. 2.9, *b*) compose up to 29% of the measured total porosity. They are no more than 0.11 μm in size and have a fissure-like form.

Examination of the laminar microstructure shows that it ranks with the microstructures having a high degree of structural element orientation. This is also confirmed by the strongly elongated ellipsoidal form of the rose of structural element orientation and by the high (A_g = 48.8%) value of the coefficient of anisotropy (Fig. 2.9, *c*).

The main distinction of the laminar microstructure from the turbulent one is the further increase in the degree of homogeneity of such a microstructure and the further decrease, together with the total porosity reduction, in the pore size. Besides, disappearance of large intermicroaggregate micropores (d = 10 to 100 μm) and interparticle ultramicropores (d < 0.1 μm) is observed in clay rocks with the laminar microstructure formed at the late catagenesis stage. The former, larger pores decrease in size under the action of high geostatic pressures and transform into small intermicroaggregate micropores ($d \sim$ 3.4 μm). The latter, interparticle ultramicropores, close up under the influence of high geostatic pressures and temperatures. This results in the «intergrowth» of fine clay particles and ultramicroaggregates.

So, an important distinguishing feature of the laminar microstructure compared with the turbulent microstructure is also the further enlargement of ultramicroaggregates and microaggregates and the increase in the degree of their orientation.

Besides the above-described turbulent and laminar microstructures, transitional microstructures may form in the continuous process of catagenesis: these are the matrix-turbulent and turbulent-laminar microstructures having intermediate characteristics.

The development of catagenesis is manifested not only in the transformation of the morphometric and geometric features of microstructures. A change in the character of interaction between structural elements and an increase in their structural cohesion are of great importance. The growth of the strength of structural bonds in such rocks is determined by several phenomena. First of all, this is connected with an increase in the number and area of contacts between structural elements in the course of rock compaction. The change in the character and energy of forces operating at the contacts is still more important. An increase in the pressure and temperature leads at first to gradual diminishing of the hydrate film thickness and then to the complete breakthrough of the film and formation of a more strong contact at a very restricted area. The contact thus formed has been termed *transitional* [51, 52].

The formation of transitional contacts is accompanied by the significant increase in the strength of clay rocks, appearance of their elastic-viscous properties, and loss of plasticity. An important feature of transitional contacts is their metastability relative to water, i.e. the capability of being hydrated and losing the properties inherent to them. The reversibility of transitional contacts is connected with the high hydration energy of the exchange cations participating in their formation, as well as with the wedging action of the adsorptionally-bound water films, which proves to be sufficient in order to overcome the cohesion forces in the small area of a transitional contact and to increase the distance between the interacting particles. One more peculiarity of clays with transitional contacts is connected with this: their high capability of swelling when moistened in the absence of the external compressing stress opposing this process.

The basis of the transitional contact formation is the formation of tight bonds due to the electrostatic attraction of negatively charged particles by the cations located between them [52] (Fig. 2.11). In the mechanism of formation, the transitional contacts are similar in many aspects to the ionic-electrostatic bonds existing inside the structures of layered silicates (e.g. micas) and ensuring the strong cohesion inside the crystals of these minerals. It is obvious that similar bonds may also arise between the clay particles with their basal surfaces closely pressed together to form a polycrystalline aggregate.

When the plane surfaces of two particles draw together at distances close to the interplanar space (0.7–1.2 nm), their diffuse layers in the zone of contact become overlapped and a specific distribution of the potential is formed (Fig. 2.12, *a*), which determines the location of cations of the particles' adsorption

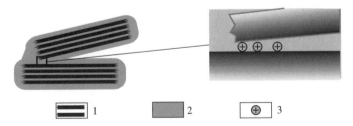

Fig. 2.11. A scheme of the transitional contact formed owing to the ionic-electrostatic interaction of clay particles: 1– clay particles, 2 – bound water, 3 – cations, forming the ionic-electrostatic bonds.

layers in the potential gap, i.e. in the center of the clearance. Such a location of positive ions leads to their interaction with both of the negatively charged surfaces and to the formation of ionic-electrostatic «bridges» between the particles (Fig. 2.12, b). With the compaction of clays and increase in the degree of particle orientation, the area of the ionic-electrostatic contact may enlarge, with its strength being concurrently increased.

Approximate calculations accomplished by the authors [52, 54] have shown that the value of the attraction force per one ionic-electrostatic «bridge» may amount to $\sim 0.3 \times 10^{-10}$ N. To form a contact that exceeds the coagulation contact in the strength of clay particle cohesion, about 1.6×10^3 ionic–electrostatic «bridges» are required. In this case, the strength of the transitional contact will be $\sim 5 \times 10^{-8}$ N and its area (within which the distance between particles is about 1.2 nm) will not exceed 1.8×10^3 nm^2, which corresponds to approximately 0.15% of the micron-size particle area.

Estimation of the upper strength limit of the transitional ionic-electrostatic contact is a matter of considerable difficulty. For this purpose, the most important feature of the transitional contact – its metastability (reversibility) – may be used. The transitional contact metastability is determined by the fact that hydration energy of the cations located in the contact zone exceeds the energy of ionic-electrostatic «bridges». As a result of contact hydration under unloading of clays, the distance between particles at the contact increases, the ionic-electrostatic attraction between them weakens, and the contact in its structure and energy becomes the coagulation-type contact.

Experiments show [52] that the reversibility of transitional contacts disappears when their strength attains $\sim 3 \times 10^{-7}$ N. At a greater strength, the transitional contacts behave as the irreversible phase contacts. It is obvious that the value $\sim 3 \times 10^{-7}$ N may be accepted as the upper strength limit of the transitional contact.

Turning back to the calculations, it is not difficult to show that the force of attraction equal to 3×10^{-7} N is reached between the micron-size particles when $\sim 10^4$ ionic–electrostatic «bridges» are formed at the contact. Assuming that each «bridge» covers an area of 1.2 nm^2, the total area of a transitional contact having

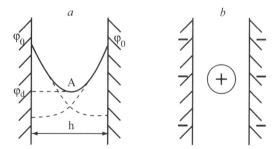

Fig. 2.12. Distribution of the like-sign potential in the clearance between two clay particles drawn together at a distance equal to the interplanar space (a), and a scheme illustrating the ionic-electrostatic interaction of particles (b).

the maximal strength will be ~$1.2 \times 10^4 \, nm^2$, which makes up about 1.5% of the particle area.

With the further compaction of clays and increase in the area on which ionic–electrostatic bonds arise, the phase contact will form. So, under certain conditions, the contacts of ionic-electrostatic nature may pass to both the coagulation and the phase contacts. On this basis, the authors have termed them the transitional-type contacts [52]. According to the calculations, the clay rocks composed of micron-size particles and having transitional contacts (the porosity being 30–25%) have a tensile strength within a 0.03–0.15 MPa range, which is in agreement with the experimental data.

Under the influence of high external pressure, the transitional contacts may arise not only between clay particles but also between sand-silt grains. In this case, however, the mechanism of transitional contact formation is different and connected with the direct press-through of hydrate films, plastic deformation of solid particles, and combined crystallization of the substance of interacting particles in the contact zone («cold welding») (Fig. 2.13).

Estimation of the external stress P_c, at which layers of adsorbed water are squeezed out of the contact zone and particles come into direct (dry) interaction, is possible if the conditions of Hertz's problem are taken into account [110]. An average value of the normal component of the external stress acting in the breakthrough zone of a hydrate adsorption film is calculated from the following expression [38]:

$$P_c = \frac{2}{\pi}\left[\frac{2}{9}\frac{E^2}{(1-\nu^2)^2}\frac{f_c}{R}\right]^{1/3} \qquad (2.1)$$

where E is Young's modulus of particles, ν – Poisson's ratio, f_c – effective external stress at an individual contact, R – radius of particles.

For the calculation, the tabulated values $E \approx 7 \times 10^{10} \, Pa$ and $\nu = 0.2$ may be taken. The value f_c is found based on the experimental data of R.K. Yusupov [111]), who studied the strength of contacts between two crystalline bodies under different conditions of their pressing together. For particles of 1–2 mm in size, the threshold value of the pressing force, at which a jump-like transition from the coagulation to more strong contacts occurs, is $5 \times 10^{-4} \, N$. By substituting these values into the given above formula one can obtain the value of

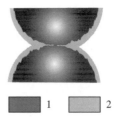

Fig. 2.13. A scheme of the transitional contact formed between sand-silt grains under the influence of external pressure: 1 – sand-silt grains, 2 – bound water.

critical stresses $P_c \approx 6 \times 10^7$ Pa, at which a transitional contact between sand-silt grains begins to form. The obtained value (60 MPa) corresponds to the data on the boundary friction [112].

The contacts thus formed have at their basis the chemical (valence) forces operating at the contacts very restricted in area. Assuming the strength of a single valence bond equal to $e^2/4\pi\varepsilon_0 b^2 \approx 10^{-9}$ N (where e is the electron charge, ε_0 – dielectric constant equal to 8.85×10^{-12} farad per meter, b – distance between atoms in a crystalline cell ~0.1 nm), one can find that for the formation of a transitional contact with the maximum strength of 3×10^{-7} N, 3×10^2 bonds are required. The area of the formed contact will be about 3 nm^2. With the number of chemical contacts being $<3 \times 10^2$, the formed transitional contact is metastable: when the external loading is removed, it is capable of hydrating with the destruction of valence bonds and transition into the coagulation contact.

With further growth of pressure and temperature at depths corresponding approximately to the middle catagenesis substage, the most strong *phase* contacts form. The basis of this type of contacts is formed by the forces of ionic-electrostatic and chemical nature. The phase contact has a high strength exceeding sometimes the strength of the contacting minerals proper. Therefore, the rocks with phase contacts are typical solid bodies deforming elastically under external loading and breaking in a brittle way when their strength limit is reached. In contrast to the transitional contacts, the phase contacts are not hydrated, which causes the resistance to water of the rocks with these contacts and the complete loss of their swelling capability.

The formation of phase contacts may proceed in several ways. When clay particles with molecularly smooth basal surfaces are closely pressed together and overlap one another over a significant area, the phase contact formation is possible at the expense of ionic–electrostatic bonds (Fig. 2.14, *a*). As has been already said, when the ionic–electrostatic bonds are formed in quantities $>2 \times 10^4$, the contact becomes water-resistant and may transform into the phase contact. This occurs with the lessening of the angle at which the particles contact. If the particles are parallel to each other, the bonds can form practically all over their area, which will allow the formation of up to 8×10^5 ionic–electrostatic «bridges». This ensures the strength of cohesion between two micron particles equal to ~10^{-5} N.

Another possible way of the phase contact formation (including contacts between non-clay sand-silt particles) is the «cold welding» of minerals owing to the forces of chemical nature (Fig. 2.14, *b*). An originally formed transitional contact passes with the further growth of pressure to the more stable phase contact. This is observed when the contact area, at which the mutual cohesion of particles occurs due to the valence forces, enlarges. When bonds due to these forces are formed in quantities $>3 \times 10^2$, the contact loses its reversibility and becomes the phase contact. The contact thus formed is similar in its nature to a portion of the grain boundary in a polycrystalline substance. Therefore, such a contact is often called a crystallization contact.

The mechanism of the crystallization contact formation is interpreted in different ways. Thus, it is believed [113] that the formation of chemical bonds

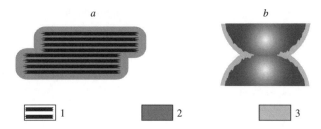

Fig. 2.14. Schemes of the phase crystallization contact formed at the expense of: *a* – forces of ionic-electrostatic nature; *b* – forces of chemical nature during plastic deformation of particles in the zone of contact («cold welding»).
1 – clay particles, 2 – sand-silt grains, 3 – bound water.

results from mutual crystallization and that the energy required for this process is liberated during the deformation of solid particles. Along with this, there exists the energy theory, according to which the formation of a phase contact requires certain energetic activation of surfaces tending to make the crystalline structure in the zone of contact amorphous.

Under the conditions of high temperatures close to the melting point of the particles, the formation of phase contacts is possible through caking. The basis of this phenomenon is formed by the processes of mutual diffusion, which extremely intensify with the growth of temperature.

The strength of an individual crystallization contact is determined by the valence bond strength ($\sim 10^{-9}$N). A phase contact with the minimal calculated strength ($\sim 4 \times 10^{-7}$N) arises when around 3×10^2 bonds are formed. Formation of a contact with the maximal strength is determined by its area depending on the mineral composition (hardness of the mineral) and the size of contacting grains. For particles with a diameter of 1 μm, the contact area may attain $\sim 10^2$ nm^2 and more, while for quartz grains with a diameter of 100 μm it is $\sim 10^4$ nm^2. Having determined the quantity of the valence bonds formed in the first and second cases, with their defectiveness taken into account, one can obtain the strength values of an individual contact equal to 10^{-6}N and 10^{-4}N, respectively.

One of the most common processes resulting in the formation of phase contacts is the cementation. Development of cementation is connected with a change in geochemical conditions within a sedimentary sequence and isolation at the contacts of grains and particles of a new phase in the form of hydroxides of silicon, calcium, iron, aluminum, and other inorganic and organic compounds (Fig. 2.15).

General regularities of the cementation process have been established in studying the hardening of astringents [114–117] and are widely used in working out the methods for the artificial consolidation of rocks. The basis of cementation is the widespread in nature process of liberation of cementing substances from the pore waters oversaturated relative to the phase being formed. The «intergrowth» of the elements of grained structure proceeds at the expense of chemical forces on the mineral-cement boundary surface. Therefore, one of the principal conditions for cementation is the chemical affinity between the cementing substance and the surface of dispersed-phase particles. The best cement for layered silicates is the polymerizing compounds capable of spontaneous formation of spatial chain and net structures and having a pronounced molecular similarity with the surface of cemented particles. For clay rocks in the presence of calcium compounds, these requirements are met by gels of silicic acids and by organic matter, which form siliceous, siliceous-carbonate, and organic cements encountered in siliceous and marly clays, as well as in schistose clays with a high content of organic matter. Besides, iron compounds and some easily soluble salts my be encountered as cements.

As shown in the studies of T.Y. Lyubimova [118], N.B. Ur'ev and N.V. Mikhaylov [119], the formation of a new phase in the contact zone proceeds at a greater rate than in the volume. This phenomenon is explained by the «organizing» influence of minerals and by the excess of surface energy owned by fine grained particles. Consequently, the cementation of mineral particles occurs in the first place along their contacts, with a rigid porous structure being formed. Along with this, the specific conditions controlling the formation of a new phase at the contacts result in the rise of additional stresses there. Therefore, the intergrowth contacts have, as a rule, excessive energy and, hence, also higher solubility as compared with the unstressed crystals formed in the volume [120].

In cementation under subaqueous conditions, it is silica that often serves as a cementing substance. According to the present concepts of the solubility of silica and its occurrence in natural waters not in the colloidal state but in the form of a true solution [73], the formation of siliceous cement in an aqueous medium occurs through the liberation of silica gels and fine-crystalline forms of silica (chalcedony, opal) at the contacts. This process is possible in the absence or insignificant presence of the magnesium, aluminum, and iron compounds in the pore solution. Otherwise, silica is spent on the formation of new clay minerals.

The processes of clay rock cementation are especially intense in the presence of calcium in the pore water and exchange complex. In this case, the formation of more complex silica compounds, such as calcium silicate or hydrosilicate, is possible [121].

Organic matter may play a considerable role in the cementation of clay rocks. It is established that under certain conditions in the process of lithogenesis, humic compounds lose their stabilizing capacity and begin to promote the formation of tight cementation bonds [121–123]. As shown by the data of infrared spectroscopy, structural X-ray and chemical analyses, the growth of the lithification degree of humic substance results in the radical change in its composition: the content of carbon that forms hydrophobic aromatic groups increases, while the amount of functional groups decreases. Concurrently, its connection with mineral particles is further strengthened: in highly lithified rocks, the organic-mineral complexes prove to be absolutely insoluble. As a result, the interaction of particles along their hydrophobic parts intensifies, with chemical bonds being formed between the chain molecules of organic compounds. This is much assisted by the prevalence in the pore solution of polyvalent cations that enter into irreversible connection with carboxyl groups and form firm bridges between the organic molecules of contacting particles.

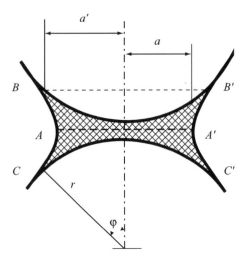

Fig. 2.15. A scheme of the phase cementation contact.

All the above-considered processes, though having different mechanisms of cementation, bring about the same result – formation of phase contacts at the expense of chemical forces and considerable increase in the strength of rocks.

The strength of the cementation contact (P_c) depends on its area and the character of destruction. When the contact is destroyed along the cement (line A–A', Fig. 2.15), its strength is estimated as follows:

$$P_c' = \sigma_c \pi a^2$$

$$(2.2)$$

where σ_c = tensile strength of the cement substance.

When the contact is destroyed along one of the grains (line B–B', Fig. 2.15), the value P_c' is found from the following expression:

$$P_c' = \sigma_m \pi (a')^2$$

$$(2.3)$$

where σ_m = tensile strength of the mineral composing the grain.

In case when the contact destruction occurs along the mineral-cement boundary (C–C', Fig. 2.15), the expression for P_c' takes the following form:

$$P_c' = \sigma_{mc} \frac{\sin\varphi}{\varphi} 2\pi r^2 (1 - \cos\varphi)$$

$$(2.4)$$

where σ_{mc} is the strength of the mineral–cement boundary zone.

A cementation contact formed by $10^2 \div 10^3$ valence bonds located in the plane of destruction may be assumed as the contact with the minimal strength. The area of such a contact is $S_c \approx (10^2 \div 10^3) \times b^2 \approx 10 \div 10^2\,\text{nm}^2$, where b is the characteristic interatomic distance in the cement substance. With a lesser number of valence bonds in the plane of destruction, the formed contact is metastable and may be destroyed under the action of the wedging pressure of liquid when the system is hydrated.

Knowing the contact area and the tensile strength of the cement substance (σ_c), one may estimate the strength of an individual cementation contact.

If we assume that σ_c for the contacts with small areas has a value close to the theoretical strength of solids, i.e. equal to $10^9\,N/m^2$, then $P'_c \approx \sigma_c \times S_c \approx 10^9 \times 10^{-16} \approx 10^{-7}\,N$. As r increases (see Fig. 2.15), the contact area grows and reaches the maximum value at $a \rightarrow r$.

Taking into account that with the growth of the contact area the defectiveness of the formed cement increases, the tensile strength of the latter is assumed to be $10^6\,N/m^2$. Proceeding from this, the maximum strength of the cementation contact destroyed along the A–A' line (see Fig. 2.15) for the systems consisting of spherical particles with a diameter of 1 and 100 μm will be $\sim 10^{-6}$ and $10^{-2}\,N$, respectively.

As can be seen from the above data, the strength of individual crystallization and cementation contacts has the same order of values varying for grains and particles from 1 to 100 μm in diameter within a range from 4×10^{-7} to $1 \times 10^{-6}\,N$ and $1 \times 10^{-2}\,N$, respectively, depending on the contact area. Therefore, in further calculations one may use an average strength of phase contacts without dividing it into the strength determined by the cementation and the crystallization contact varieties.

The calculations show that fine grained systems with porosity of 5–15% having phase contacts and an average size of structural elements within the range of 2–3 μm have the tensile strength amounting to 4–6 MPa, which corresponds to the average strength of highly lithified clay rocks from the middle and lower subzones of catagenesis.

2.2.3. METAGENESIS

Further transformation of clay rocks in the lower parts of the stratisphere proceeds under the influence of metagenesis processes. These processes are close in their character to the initial stages of metamorphism. Metagenesis manifests itself within the lower and middle structural stages of geosynclines at depths from 5 to 15 km with temeperatures attaining 200–300 °C and pressures no less than 200 MPa. Stress may play a considerable role in metagenesis. Under the platform conditions, metagenetically transformed sequences are encountered among ancient (Precambrian) rocks, where they have the appearance similar to that of rocks altered by the low-temperature regional metamorphism.

The compaction of rocks comes on the whole to an end by the beginning of metagenesis. Their transformation proceeds mainly at the expense of physicochemical processes. The widespread processes are dissolution and regeneration, under whose influence the bulk of the rock become recrystallized, with polycrystalline overgrowths of platy crystals being formed instead of the leaflike varieties. As a result, mudstones and siltstones turn into shales, slates, and phyllite-like slates.

Hydromicatization of authigenous and clastic clay minerals intensifies, with high-temperature hydromica polytypes being formed. Among clay minerals, the association represented by dioctahedral hydromica $2M_1$, sericite, chlorite, sometimes dickite and pyrophyllite is widespread (see Fig. 2.1).

Recrystallization under the high gravity and stress conditions results in the profound structural alterations of clay rocks. There arise schistosity oriented, as a rule, at some angle to the bedding, and flow cleavage with parallel orientation of platy crystals.

The morphometric features of the original deposit become completely leveled, and many metamorphosed clay rocks acquire the blastic microstructure composed of microlayers and platy crystals up to 3–5 μm thick, well oriented along the bedding (Fig. 2.16, *a*).

The rocks have the highest degree of structural element orientation amounting sometimes to 75% (Fig. 2.16, *b*). Their characteristic features are low porosity (1–2%), the absence of interparticle ultramicropores. The pore space is mainly composed of small intercrystalline micropores of a fissure-like form.

The intensified crystallization leads to the increase in the number of crystallization contacts and decrease in the cementation-type contacts. At the same time, heterogeneity of structural bonds grows resulting from the development of rock schistosity and cleavage.

2.3.   BOUNDARIES BETWEEN THE STAGES AND SUBSTAGES OF LITHOGENESIS

Delineation of boundaries between individual stages of lithogenesis remains up to now one of the most actual problems in lithology. Thus for example, N.M. Strakhov [77, 78] believes that the boundary between the stages of diagenesis and catagenesis may vary in depth depending on specific conditions from 10–50 to 200–300 m. N.V. Logvinenko [81] notes that catagenetic alterations begin at a temperature of 30–50 °C and end at 150–200 °C. The uniaxial pressure values corresponding to these temperatures are equal to 10–20 and 150–200 MPa, respectively. G. Müller [89] defines the lower boundary of the diagenesis stage at a depth of no more than 500 m. At the same time, he proposes to bring down the lower boundary of the catagenesis stage to the depths of 5–10 km. G. Larsen and G. Chilingar [124] consider it possible to define the lower boundaries of the early and late diagenesis at the depths of 10 and 400 m respectively. F.S. Aliev [125] defines the boundary of the early diagenesis substage in clay sediments of the Baku archipelago at a depth of 10 m and that of the late diagenesis substage at a depth of 60 m. I.G. Korobanova [123], who has studied a continuous section of the Alyaty-marine area in the Baku archipelago region of the Caspian Sea to the depth of 1207 m, distinguishes in the upper part of the section the early diagenesis subzone extending to the depth of 8 m and the late diagenesis subzone extending to depths of 70–80 m. According to her data, the lower boundary of catagenesis lies beyond the studied depth interval (1200 m). A.S. Polyakov [126], based on the study of the physical properties and microstructures of clay sediments from the Baku

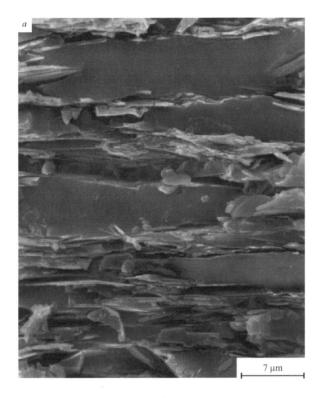

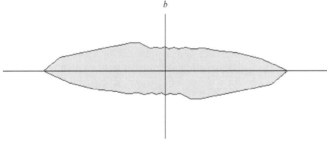

Fig. 2.16. The blastic microstructure of a metamorphosed clay rocks (Sochi area) (*a*); the rose of orientation of structural elements (*b*).

archipelago and from the deep-water part of the Black Sea, draws the boundary between diagenesis and catagenesis at a depth of 50–200 m.

The main reason for the difference of opinion concerning the boundary delineation is, firstly, the influence of depositional environments on the development of lithogenesis and, secondly, the absence among different schools of thought in lithology of a uniform concept about diagenetic indices and criteria for the differentiation of individual lithogenesis stages. Rock porosity is most often used as the principal criterion. Indeed, the available data obtained in laboratory experiments [95, 127] and in studying the natural compaction of clays [123, 128–133] evidence for the existence of a regular tendency in the process of rock compaction depending on the depth of burial. At the same time, all the researchers acknowledge that the degree of post-sedimentation changes in the porosity of rocks depends to a large extent on their chemical-mineral composition, grain size distribution, content of organic matter, physicochemical conditions of the environment (composition of exchange cations, pH, Eh, composition and salinity of pore solutions), age, and other factors. So, in each specific case, compaction of clay deposits under the influence of geostatic pressure develops in a different way. It should be noted that the laboratory data, although supporting the general regularity in the development of compaction with the growth of loading, do not correspond in full to all the thermodynamic and physicochemical conditions of the natural process and, consequently, can not serve as a reliable foundation for establishing the dependence between the rock porosity and degree of lithification.

The above evidences that clay rocks cannot be considered as a simple mechanical model allowing one to establish the history of their formation from the porosity indices. In delineating the boundaries of diagenesis and catagenesis, not only the degree of rock compaction, but also a number of other indices must be taken into account. The structural bonds may be one of such indices.

If one assumes that the stages of lithogenesis are related to the qualitative transformations of the composition and condition of clay deposits, then the transition from one stage to the other must be accompanied by a noticeable change in the strength and deformation properties of clays and clay rocks. These changes, as will be shown in the following chapters, are determined first of all by the transition of one predominant type of contacts into another one, i.e. by the transformation of the energy type of structure. Knowing the values of pressures and temperatures corresponding to the formation of one or other type of contact, it is possible to define the depths, at which noticeable changes of clay properties are expected, and to connect these depths with the boundaries of different lithogenesis zones.

Another important index of the clay deposit condition, closely associated with the character of contact interactions, is the content in clays of different categories of bound water and its removal under certain pressure-temperature conditions. In the process of lithogenesis, various categories of bound water are gradually removed from clays: free water is squeezed out first, after which, as the temperature and pressure grow, the loosely bound and then the tightly bound (adsorption) water is gradually squeezed out . The loss of free and physically bound water is accompanied by the transformation of clay structure and properties, including the character of contact interactions.

The clay minerals themselves bear an important information on the degree of alteration in lithification. The transformations of clay minerals connected with specific pressure-temperature and physicochemical conditions and the associations formed by these minerals may provide valuable evidences for different stages of lithogenesis.

At last, data on the organic matter alteration and degree of coal metamorphism obtained from vitrinite reflectance values can be interpreted as closely associated with the stages of lithogenesis development.

The aforesaid evidences for the versatile character of the lithogenesis process. This means that in studying the transformations of clay sediments, one should proceed from the combined analysis of a number of clay characteristics, which are to some or other extent indicative of the stages and substages differentiated in the process of lithogenesis. The following may be used as such indices: rock porosity, clay mineral paragens, degree of coal metamorphism, character of structural bonds, and content of different categories of water. With due regard for the above considerations, one may turn to the substantiation of boundaries between separate stages of lithogenesis.

2.3.1.   BOUNDARIES BETWEEN THE DIAGENESIS SUBSTAGES AND CATAGENESIS

Proceeding from the character of structural bonds and consistence of sediments, the diagenesis stage may be subdivided into two substages: the early and the late one (Table 2.1). The boundary between the substages is drawn at a depth of 8–15 m.[1] Sediments at the lower boundary of the *early diagenesis substage* are characterized by extremely weak structural bonds determined by the development of distant and close coagulation contacts. When the structure is disturbed, the sediments acquire fluid consistence. The sediment porosity at this substage remains not lower than 60–65%. A considerable quantity of free water is preserved in pores of the sediment. Therefore, the water content of the sediments exceeds that of the liquid limit.

The lower boundary of the *late diagenesis substage*, which is at the same time the boundary between the stages of diagenesis and catagenesis, corresponds to the transition from a sediment to a consolidated clay deposit. No doubt, that such a transition is gradual and there is no definite interface. So, the boundary between diagenesis and catagenesis varies within a considerable depth range.

Proceeding from the available data, this boundary may be suggested to lie at depths of 80–90 m for shallow-water clay sediments, while for fine grained sediments deposited in deep-water environments it sinks to depths ranging from 150–300 to 500 m (Table 2.1). The geostatic pressures of 2–10 MPa and temperatures not exceeding 15–20 °C correspond to these depths. The porosity of clay deposits at the boundary is 35–45%.

The composition of clay minerals shows no noticeable changes with the transition from diagenesis to catagenesis. Organic matter is gradually transformed from turf and sapropel into brown coal. Water in the free state may be preserved only in individual large pores. The physically bound (osmotic and adsorption) water is prevalent. The water content is equal to, or slightly less than, that of the liquid limit. The structural bonds are mainly determined by molecular forces operating at the contacts through the hydrate films of bound water. The close coagulation contacts are dominant.

---

[1]Here and further, «depth» means the thickness of the overlying sediment or rock sequence, and not the sedimentation basin depth.

Table 2.1. Stages and substages of lithogenesis of clay sediments and rocks.

Lithogenesis stages	Lithogenesis substages	Depth to the lower boundary, m	Pressure at the lower boundary, MPa	Temperature at the lower boundary, °C	Clay mineral association	Vitrinite reflectance, $10R^a$	Total porosity at the lower boundary, %	Water content at the lower boundary, %	Consistence	Prevailing type of contacts
Diagenesis	Early	8–15	0,15	10–15	Montmorillonite, illite, kaolinite, mixed-layer minerals		60–75	$W \gg W_L$	Latent-fluid	Distant and close coagulation
	Late	80–300 (500)	2–10	15–20	The same as at the previous substage		35–45	$W \leqslant W_L$	Plastic	Close coagulation
Catagenesis	Early	900–1800 (2000)	20–30	50–60	Illite, montmorillonite, mixed-layer minerals, kaolinite	65–75	16–25	$W \geqslant W_P$	Semi-solid	Coagulation and transitional
	Middle	2100–3600	60–80	80–100	Illite, mixed-layer minerals, chlorite	75–80	4–12	$W \leqslant W_{mh}$	Solid	Transitional and phase cementation
	Late	2600–5000	120–200	150–200	Illite, chlorite	80–90	2–4	$W \ll W_{mh}$	Solid	Phase crystallization and cementation
Metagenesis		10000–15000	>200	>200	Dioctahedration of hydromica; sericite, chlorite		1–2	$W \ll W_{mh}$	Solid	Phase crystallization

2.3.2. BOUNDARIES BETWEEN THE CATAGENESIS SUBSTAGES AND METAGENESIS

In spite of the existing specificity in the catagenesis development under various geological conditions, there are general regularities, in correspondence with which clay deposits of different initial compositions, genesis, and ages acquire in the course of catagenesis some characteristic features allowing their subdivision according to the degree of catagenetic transformations. This makes it possible to distinguish in catagenesis, rather conventionally, three substages: early, middle, and late catagenesis [79, 92, 134].

As has been already noted, catagenetic processes depend on the composition and homogeneity of clay deposit sequences, connected with which are the rate of pore solution squeezing-out, distribution of geostatic pressure, intensity of geochemical processes, total time of lithogenetic transformations. Therefore, the lower boundary of the *early catagenesis* may vary within a significant range: in the coarser and heterogeneous sediments, it is reached at depths of 900–1000 m, while in the fine grained homogeneous clays it may sink to depths as great as 1800 m, in rare cases to 2000 m (Table 2.1). The pressure corresponding to this depth is 20–30 MPa and the temperature is 50–60 °C. The clay porosity decreases to 16–25%. The deposits are represented by dense, sometimes weakly indurated clays of semi-solid consistence. The association of clay minerals undergoes insignificant changes in comparison with late diagenesis. The minerals prevailing in the clay fraction are of various composition, mainly allothigenic, beginning from montmorillonite and mixed-layer varieties and ending by hydromica and kaolinite. The processes of hydromicatization with the formation of mixed-layered phases are little developed. Therefore, the early catagenesis substage is often called the subzone of unchanged (or little changed) clay substance.

In the pore moisture, free water is practically absent, bound (osmotic) water is being intensely squeezed out. By the end of the substage, mainly adsorptionally bound water remains in the clay. In accordance with this, water content of the deposit becomes equal to, or slightly higher than, that of the lower plastic limit.

The increase in clay density and removal of part of the bound water promote the strengthening of structural bonds at the expense of increase in the number of contacts between structural elements and rise of transitional contacts of ionic–electrostatic nature. Concurrently, the processes of cementation and formation of transitional contacts of chemical (cementation) nature begin to develop.

Among the fossil coals, a transition from brown to long-flame coals is registered. The vitrinite reflectance values are not greater than 65–75.

The boundary between the *middle* and the *late catagenesis* substages corresponds to depths of 2100–3600 m. The lithostatic pressure at these depths is 60–80 MPa, temperature attains 80–100 °C. The porosity of rocks does not exceed 4–12%. The rocks are represented by well-compacted indurated clays passing at the base of the subzone to shales. A change of the clay mineral association is observed resulting from the transformation of montmorillonite into the mixed-layer phase, illite, and chlorite. The long-flame coals change into the gas fat coals. The vitrinite reflectance increases to 75–80. The rock water content is a little lower than, or close to, the maximal hygroscopic value.

The concentration of salts in the pore solution grows due to the enhanced protonization and dissolving power of pore water. The dissolved substance is removed from the rock together with the squeezed-out water or serves as a source for the crystallization of new minerals with more perfect structure and enlarged crystals. The compaction and growing together of clay minerals begin inside the aggregates to form large clay polycrystalline overgrowths character-istic of argillites.

At a depth below the 65–70 °C isotherm, with the appearance of highly aggressive pore moisture, the cementation gives place to the opposite process – dissolution of cement and decrease in the structural strength of the rock. However, the subsequent processes of recrystallization result in the secondary cementation of the rock and new strengthening of its structure.

Determination of the *lower boundary of the catagenesis zone* is usually connected with the completion of processes controlling the formation of sedi-mentary rock sequences. Accurate determination of the boundary between cata-genesis and metagenesis is a difficult problem that remains unsolved up to now. This is explained by the low rate of catagenesis that continues during many tens and even hundreds million years. Therefore, the lower boundary of catagenesis in the present sedimentation basins has not been reached even by the deepest holes. Neither has it been reached from the world ocean bed. Hence, the accu-rate definition of the catagenesis boundary becomes highly problematic. Following the statement of G. Müller [89], it may lie at a depth of 5–10 km. The temperature at this boundary is suggested to attain 150–200 °C and geostatic pressure – 120–200 MPa. The porosity of rocks does not exceed 2–4% (Table 2.1). The clay rocks are represented by argillites and siltstones, with hydromicas and chlorite prevailing in their clay mineral association. The water content of the rocks is several per cent; it is determined by the presence of remnant (not squeezed-out) pore moisture formed chiefly at the expense of hydromicatiza-tion of mixed-layer minerals. The contacts between structural elements are of the crystallization and cementation types.

Conditions for the development of recrystallization processes are preserved, resulting in the formation of minerals more stable in the given thermodynamic environment, the overgrowth and enlargement of their crystals. Concurrently, sec-ondary cementation and rock strengthening proceed. The clay matter forms large polycrystalline overgrowths of a flattened form, whose orientation along the bed-ding determines the schistosity of argillites and, to a lesser extent, of siltstones.

3

Formation of the Properties of Clay Seals in Lithogenesis

The properties of clay rocks depend on their composition (grain size distribution, the composition of minerals, organic matter, and pore waters), consistency (density and water content), structural features, and on the pressure-temperature conditions (stressed state and temperature) of their occurrence. All the above factors are determined by the conditions of clay sediment formation and by all the following processes that develop in clay sediments in the course of lithogenesis. Therefore, the properties of clay seals progressively change during the geologic history, the nature of these changes being determined by the characteristic features of lithogenesis processes.

3.1. DENSITY AND POROSITY

Density and porosity are the most important characteristics of the rock physical properties. Both parameters undergo significant transformations in the course of lithogenesis and are functionally interconnected. As, to estimate porosity, one must know indices characterizing rock density, the latter will be considered first.

3.1.1. DENSITY

Density is characterized by three indices: density of rock (volume density), density of its mineral part (mineral density), and density of its matrix (dry rock density).

The most commonly used index is *rock density* (ρ) which is characterized by the mass of a volume unit of a rock having natural (undisturbed) structure and water content and is expressed in g/cm^3 or t/m^3. The *mineral part density* (ρ_m) gives the average density value of minerals composing the rock. The *dry rock density* (ρ_d) index corresponds to the mass of a volume unit of dry rock with an undisturbed structure. The values of ρ and ρ_m are found experimentally in laboratory, while the ρ_d value is calculated according to the formula:

$$\rho_d = \frac{\rho}{1 + 0.01W} \tag{3.1}$$

where W = weight water content of rock in %.

All the three density indices change in the course of lithogenesis, but the factors determining these changes are different for each of them. The density (ρ) values are determined by the changes in porosity, water content, and mineral part density of the rock. The ρ_d values depend on the porosity and density of the mineral part of the rock, while the ρ_m value depends only on the density of minerals and solid organic matter that compose the rock.

Among the density indices, the ρ_d and ρ values undergo the greatest changes in the course of lithogenesis. They grow with the compaction of rocks and reduction of their porosity. As the subaqueous lithogenesis occurs under the conditions of complete water saturation of rocks, the increase in their density proceeds concurrently with decrease in their water content.

An analysis of clay deposit density values within different zones of lithogenesis allows one to understand some general regularities in the change of density with depth. Interesting results have been obtained by I.G. Korobanova et al. [135] for the Neogene-Quaternary clay sediments of the Caspian Sea. As is seen from the plot (Fig. 3.1), the values of sediment density are 1.62–2.08 g/cm^3 in the subzone of early catagenesis which, according to the authors, begins at the depth of 60 m.

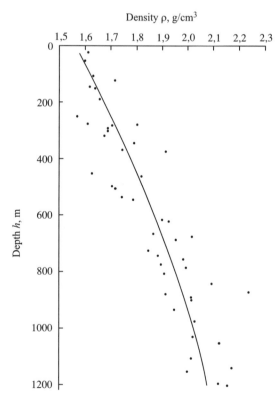

Fig. 3.1. Change in the density (ρ) of clay sediments with depth (h) for the Alyaty-Marine area of the Caspian Sea [135].

Data on clay rock densities in deeper subzones are contained in the works by A.A. Khanin et al. [136], G.Eh. Prozorovich [107], N.B. Vassoevich [108], V.M. Lazareva [137, 138], O.A. Martirosova and V.V. Bakhtin [139], I.D. Zkhus and V.V. Bakhtin [103], and others.

Abundant data on the density of clay rocks in the Western Pre-Caucasus region has been obtained by A.A. Khanin et al. [136], A.A. Khanin [140]. They have established that rock densities in the depth interval of 1000–2000 m range from 2.05 to 2.34 g/cm^3, at the 3000 m depth the average rock density is 2.55 g/cm^3, at 4000 m it is 2.65 g/cm^3, and at 4500 m –2.70 g/cm^3. It has been noted that the scattering of clay density values decreases with depth. With all other conditions being the same, the clays of kaolinite and partially hydromica composition are being compacted more intensely than those containing mixed-layer minerals and montmorillonite.

The dry rock density (ρ_d) is functionally related to the volume density and water content of the rock and is found, as has been said earlier, by calculation using Formula (3.1). A number of authors have performed the calculation of the ρ_d value and obtained the plots of its change with depth. Thus, N.B. Vassoevich [108], having generalized the density-depth dependence, obtained the ρ_d values varying from 1.40 g/cm^3 in the diagenesis zone to 2.65 g/cm^3 in the late catagenesis subzone. Figure 3.2 shows the dry density vs. depth plots for the

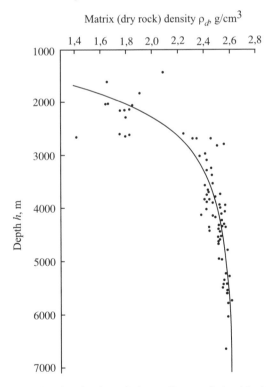

Fig. 3.2. Change in the dry density of clay sediments (ρ_d) with depth (h) for the Northern Peri-Caspian region [141].

Northern Peri-Caspian region, based on drilling data from the deep Aralsor well [141]. As can be seen on the plot, the matrix density of rocks in the Northern Peri-Caspian region is 1.4–$1.6 \, g/cm^3$ at the early catagenesis substage, 1.6–$2.4 \, g/cm^3$ at the middle, and 2.4–$2.6 \, g/cm^3$ at the late substage.

The density of the mineral part of rock, ρ_m, undergoes the least changes in the course of lithogenesis. Its values are mainly determined by the initial mineral composition of clay sediments, which is formed during sedimentogenesis; later they undergo insignificant changes connected with the processes of mineral solution and neocrystallization in the course of lithogenesis.

In the zone of diagenesis, the density of the mineral part of rock remains practically unchanged. It is determined by the mineral composition of the clay and clastic (sand and silt) parts of the sediment, as well as by the amount of organic matter. Its values range from 2.63 to $2.78 \, g/cm^3$ averaging $2.75 \, g/cm^3$. The ρ_m value is considerably influenced by the content in the sediment of organic matter, whose density is much lower than that of the main minerals and is 1.20–$1.40 \, g/cm^3$ for humus. The density of the mineral part of sediment can also decrease with increasing contents of montmorillonite and mixed-layer minerals with swelling layers. The hydration of the inter-layer space and increase in volume (swelling) reduce the density of these minerals to 2.2–$2.5 \, g/cm^3$. Some increase of ρ_m may occur in the zone of diagenesis down the section in connection with the processes of organic matter alteration and formation, under certain conditions, of sulfides (pyrite, marcasite), iron hydroxides, gypsum, and carbonates having higher mineral density values than the principal mineral components of the sediment.

In the subzone of early catagenesis, the ρ_m values remain largely within the range of 2.65–$2.80 \, g/cm^3$ and do not undergo any regular changes. This is explained by the preservation in this subzone of the same principal paragens of clay and clastic minerals as in the higher subzone.

Beginning with the middle catagenesis and ending in metagenesis, the processes of clay mineral alteration and non-layered silicate recrystallization in clay rock sequences intensify, with new paragens of authigenic minerals being formed. Montmorillonite and mixed-layer varieties become partly substituted by hydromicas, sericite and muskovite come into being, chlorite increases in content. An increase in the contents of higher-density minerals leads to some increase in the density of the rock mineral part as a whole. The ρ_m value averages $2.74 \, g/cm^3$ in the middle catagenesis subzone and $2.76 \, g/cm^3$ in the late catagenesis subzone.

3.1.2. POROSITY

Porosity is the most important characteristic of rock compaction. Therefore, much attention is given to estimating the porosity values when studying the alteration of rocks in the course of lithogenesis.

As has been already said in Chapter I, several porosity types are usually differentiated: total, open, closed, and effective porosity. It is the total and the open porosity that are most often examined in clay deposit studies. The former is found by calculation, while the open porosity is determined using experimental methods.

To characterize the *total porosity*, its two indices are used: porosity and void ratio. The former is determined by the pore volume to total rock volume ratio in percent and is calculated using the formula:

$$n = \frac{\rho_m - \rho_d}{\rho_m} \times 100\% \qquad (3.2)$$

where ρ_m and ρ_d are mineral density and dry rock density, respectively.

The void ratio is a dimensionless value and is characterized by the ratio of pore volume to the volume of solid (mineral) elements of the rock. The void ratio value is found using the following expression:

$$e = \frac{\rho_m - \rho_d}{\rho_d} \qquad (3.3)$$

There is a direct relationship between the two indices of total porosity:

$$e = \frac{n}{1 - n} \qquad (3.4)$$

The *open porosity* is determined experimentally under laboratory conditions using various methods [142], the most common among which is the mercury method of porosity metering. The method is based on the filling of pore space with mercury injected into the rock under high pressure. Owing to the low wettability of clay minerals with mercury, the latter is able to penetrate under pressure even into the ultramicropores having sizes as small as 0.01 μm. The size of pores being filled by mercury is related to the external pressure acting on the mercury as follows:

$$P_c = 2\sigma\cos\theta / r \qquad (3.5)$$

where P_c – capillary pressure counterbalanced by the external pressure transmitted to the mercury, σ – surface tension of mercury on the mineral-mercury interface, r – average equivalent pore radius, θ – angle of mineral wetting with mercury.

Having determined the quantity of mercury used for pore filling under the action of gradually increasing external pressure, one can obtain the pore volume distribution as a function of pore radius.

The total porosity values found by calculation are often close to the open porosity values determined using the mercury method of porosity metering. Some difference between these values is encountered in clay rocks showing an increase in the amount of closed micropores inaccessible for mercury. In such cases, the open porosity values remain always lower compared to the total porosity.

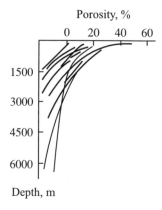

Fig. 3.3. Dependence of porosity on the depth of sediment burial [146].

In estimating the porosity, it is necessary to take into account the elastic decompaction of rocks during their rise from great depths. In plastic clays, the elastic strains are little pronounced and have no noticeable influence on the true value of porosity. As the clay cementation increases and microfractures appear, the elastic release grows and may cause some error in the porosity value estimate. According to L.M. Marmorshtein [142], relative changes in porosity in such cases may attain 30%.

The changes in clay sediment porosity in the course of lithogenesis are described by many researchers. Examples of diagenetic compaction of muds are given in the works by N.M. Strakhov [143], V. Engel'gardt [144], G. Müller [89], I.G. Korobanova [123], and many others. These authors have shown that the initial porosity of sediments is as high as 85–95%, at a depth of 2–3 m from the sediment surface it is reduced to 70–75%, while at a depth of the early diagenesis boundary (8–15 m) the porosity is 60–75% or less. The coarser sediments of the shallow-water shelf and littoral zones of inner seas and oceans are compacted most intensely, and the fine grained sediments of the deep-water zones of sedimentation basins are compacted at the slowest rate.

The catagenetic alteration of clay porosity is described in the works by N.B. Vassoevich [108], J. Weller [130], R. Meade [90], A.A. Khanin et al. [136], Y.V. Mukhin [96]; G.Eh. Prozorovich [107], O.A. Martirosova and V.V. Bakhtin [139], I.G. Korobanova [123]; A.S. Polyakov et al. [145], G. Chillingarian and X. Rieke [146], and many other authors.

Figure 3.3 shows data obtained by G. Chillingarian and X. Rieke [146] for different oil- and gas-bearing basins of the Atlantic and Pacific regions. Similar data have been obtained for a number of oil- and gas-bearing regions of the former Soviet Union (Fig. 3.4).

Abundant data have been generalized by G.Eh. Prozorovich [107] in studying the dependence of open porosity on the depth for the central and northern areas of West Siberia (Fig. 3.5).

As is seen from the given plots, there is a considerable scattering of porosity values for the same depths. This circumstance is quite understandable if

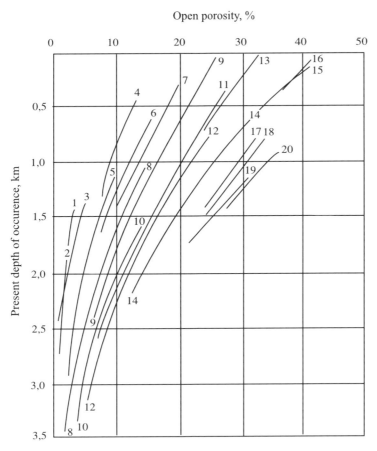

Fig. 3.4. Dependence of the open porosity on the depth of sediment burial for different oil and gas fields [109]: 1 – Troptun, 2 – Lopatinskaya, 3 – Polyanskaya, 4 – Sredniye Lyagry, 5 – South Gyrgylan'i, 6 – Mukhto, 7 – Ehrri, 8 – South Komulan, 9 – Katangly, 10 – Astrakhanskaya, 11 – East Ossoy, 12 – Uzlovaya, 13 – Severo-Dolinskaya, 14 – Yuzhno-Dolinskaya, 15 – Malinovskaya, 16 – Zelenodol'skaya, 17 – Lugovskaya, 18 – Yuzhno-Lugovskaya, 19 – Yuzhno-Lugovskaya, 20 – Tarnayskaya.

one takes into consideration the great diversity of sediment compositions and the fact that there are no basins, whose environments of sediment formation and alteration are completely identical.

According to the results obtained on the porosity change with depth, a number of authors distinguish several stages in the continuous process of lithogenetic compaction of clay sediments. D. Hager and D. Handin [147] distinguish four compaction stages: the stage of mechanical compaction (porosity 75–95%), the stage of dewatering (35–75%), the stage of mechanical deformation (10–35%), and the stage of recrystallization (<10%). N.B. Vassoevich [148] has differentiated the stages of free compaction (initial porosity 45–50%), impeded compaction (35–45%), and highly impeded compaction (<8%).

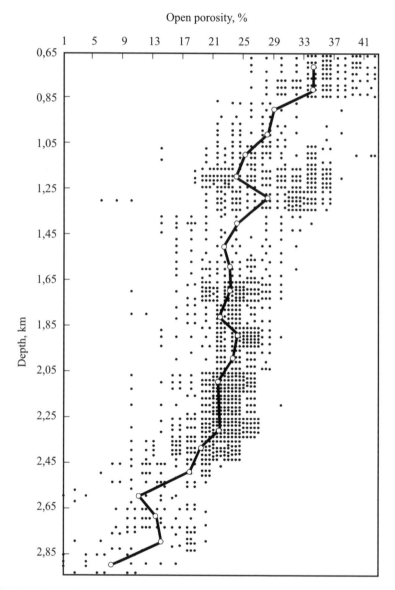

Fig. 3.5. Dependence of the open porosity of arenaceous-aleurolitic rocks on their depth of occurrence (central and northern regions of the West-Siberian oil- and gas-bearing basin) [107].

Proceeding from the published data and the present authors' own experience, the principal regularities in the change of pore space in clay sediments will be described below. It is noteworthy, that most of the researchers who studied the transformation of pore space in the course of lithogenesis, used to estimate the change in the total porosity of rocks, while paying less attention to the transformation of the size, shape, and relation of different genetic types of pores. However, an analysis of the morphometric indices of pores and the dynamics of transformation of their different genetic types gives extremely valuable information for understanding the mechanism of pore space change and for estimating the screening properties of clays. Therefore, when describing the rock porosity change in lithogenesis, we shall pay attention not only to the porosity values but also to the pore space structure.

As has been already said in Chapter 1, the sediment formed in the course of sedimentogenesis is characterized by a honeycomb microstructure having extremely high porosity, whose values grow with a decrease in particle size. In this sediment, practically all genetic varieties of pores are represented – from interparticle to intergranular.

A quantitative analysis of the pore space structure has been performed following the authors' technique [69] using the computer processing of scanning electron-microscopic (SEM) images of structures in the young sediments of the Black and Caspian Seas. It has shown the common development in them of large isometric intermicroaggregate micropores 10–15 μm in size, as well as of small intermicroaggregate pores with an average equivalent diameter close to 5 μm. The sum of these pores amounts to 97% of the total measured porosity. Among other pores, fine interultramicroaggregate pores (diameter 0.2 μm) and anisometric interparticle ultramicropores (0.06 μm) are developed.

Under the influence of at first physicochemical processes and then the increasing geostatic pressure, the sediments begin to be compacted and reduce their porosity. By the end of the early substage of diagenesis their porosity is reduced to 60–75% on the average, and at the end of the diagenesis stage it equals 35–45%. The predominance of loose distant and close coagulation contacts in the sediments determines the high mobility and compactibility of their structure. The compaction proceeds mainly due to the mutual displacements of microaggregates and destruction of large intermicroaggregate micropores which disintegrate into smaller micropores. Towards the base of the diagenesis zone, the structure of the sediment pore space becomes more homogenous, large (10–100 μm) and part of small (1–10 μm) intermicroaggregate micropores practically disappear, fine (0.1–1 μm) intermicroaggregate pores increase in content. As a weak orientation of structural elements develops, the number of anisometric intermicroaggregate micropores increases. In the total pore balance, the intermicroaggregate micropores keep to be predominant, as before.

When passing to the catagenesis zone, the processes of pore space transformation have approximately the same trend as at the end of diagenesis. In the early catagenesis subzone, the mobility of the structure remains high owing to the partially preserved coagulation contacts. Therefore, as the geostatic pressure grows, intense compaction and pore space reduction go on at the expense of mutual displacement of structural elements and their more close packing. Towards the end of the early catagenesis substage, the sediments lose a considerable volume of their pore space which is reduced to 16–25% on the average.

At this substage of lithogenesis, the pore space structure undergoes noticeable changes. According to the quantitative analysis data, the bulk of the pore space volume is made up of the intermicroaggregate pores that account for about 93% of porosity. Among these pores, small (average equivalent diameter 3 μm) and fine (0.4 μm) micropores are predominantly distributed. The pores are anisometric and isometric in shape, the latter being predominant. The intramicroaggregate porosity represented by fine interultramicroaggregate (0.4 μm) and interparticle (0.06 μm) ultramicropores is of subordinate significance. Both types of the intramicroaggregate pores are anisometric in shape.

At the end of the early catagenesis substage, the mechanism of clay compaction becomes significantly changed. This is connected with the beginning of cementation in rocks and the predominance in them of tight, breaking in a brittle way transitional and phase contacts. The compaction proceeds mainly through stress accumulation at structural heterogeneities – the large pores and microaggregates, resulting in their destruction. The elevated temperature considerably plastisizes the rock, so that its destruction remains rather plastic than brittle in character. However, the increasing strength of structural bonds slows down the process of clay compaction. In spite of the impeded conditions for the mutual displacement and packing of structural elements, porosity in the middle catagenesis subzone continues to be reduced and is on the order of 4–12% at the lower boundary of the subzone.

The pore space structure becomes more homogeneous. Destruction of the largest micropores and microaggregates leads to the predominance of rather homogenous in size, fine intermicroaggregate micropores with an average equivalent diameter of 0.4–0.6 μm. Large and small micropores are preserved only at places of pelitic and sand grain accumulation. The anisometry of pores becomes more pronounced due to the increased level of structural element orientation. The latter causes also a decrease in size of the interultramicroaggregate and interparticle pores that pass completely to the rank of ultramicropores.

At the end of the middle and beginning of the late catagenesis substage, the formation of the pore space and its structure begins to be affected by the processes of rock matter recrystallization leading, with the beginning of argillization, to the mutual intergrowth of clay particles and transformation of microaggregates into polycrystalline overgrowths of a platy shape. The enlargement of structural elements leads again to the formation of stress concentrators causing the local breaks of the rock structure continuity. The break of structure occurs along the most weakened zones, which are the boundaries of polycrystalline overgrowths. The arising breaks initiate the formation of microfractures in argillites.

The above indicates that at the boundary of the middle and the late catagenesis substages, a change in the mechanism of pore space formation takes place again: along with the continuing process of retarded compaction, the processes of crystallization and recrystallization of rock matter begin to play an important role.

The combined action of the compaction and recrystallization processes results in the further reduction of porosity: by the moment of the accomplishment of catagenesis and beginning of metagenesis, its value becomes no more than 2–4%. The pore space structure undergoes a radical change. The interparticle and interultramicroaggregate ultrapores practically disappear. The number of inter-microaggregate micropores considerably decreases. Instead of them, the microfractures form, which make the main contribution to the void volume. The largest microfractures do not exceed 0.4–0.6 μm in width and 3–10 μm in length.

At the stage of metagenesis, the mechanism of pore space transformation becomes of the pure recrystallization type. The restructuring of the whole mineral substance begins in the rock with the formation of new paragens of authigenic minerals, transformation of the structure, and progressive development of schistosity and cleavage. Microfracturing becomes the main type of void space in the rocks, being responsible for their permeability. The development of microfracturing has its peculiarities. As the microfractures form, they increase in size up to some critical values, after which they stop growing and begin to be «healed» by the newly formed authigenic matter, most often of silicate or carbonate composition.

Proceeding from the above, the general picture of the pore space formation in clays in the course of lithogenesis and the differentiation of separate stages in this transformation can be made more precise. The differentiation of the stages should be based on the change in the character of structural bonds with depth, as well as on the development of recrystallization and authigenous neocrystallization processes which determine the mechanism and the rate of pore space transformation. The geostatic pressure, which is in the focus of many authors' attention, effectively influences the clay sediments only up to the moment when strong structural bonds have formed in them. With the beginning of induration and cementation, the gravitational compaction slows down significantly, and with the transition of rocks to the late catagenesis subzone and especially to the metagenesis stage, it ceases to play the leading part in compaction. Further transformation of porosity proceeds mainly through the processes of recrystallization and neocrystallization.

The aforesaid is substantiated by the experimental laboratory data on the compaction of clay sediments under high pressures. Thus, in the experiments by V.D. Lomtadze [95], the compaction of Cambrian clay under the loading of up to 300 MPa has allowed the reduction of its porosity to a minimal value of 7%, while under natural conditions the porosity of clay rocks under such loads is no more than 1–2%. It should be noted that laboratory compaction of clays proceeds under more favorable conditions than in a natural environment, as it is not accompanied by the formation of cementation contacts. It is interesting that, when compacting a paste of the same Cambrian clays, the porosity of 7% has been achieved under the loading of 100 MPa.

Thus, several stages may be distinguished in the geostatic compaction and porosity change of clay sediments (see Table 3.1):

The first stage – that of free geostatic compaction developing in diagenesis and at the early substage of catagenesis. Its characteristic feature is the predominance in the clay deposit of the coagulation contacts which provide, owing to the presence of bound water films, the high mobility of the structure and the

Table 3.1. Alteration of the condition and properties of clay sediments in subaqueous lithogenesis.

Lithogenesis	Substages	Predominant contact type	Compaction Stage	Porosity, %	Dehydration Stage	Water content, %
Diagenesis	Early	Distant and close coagulation	Free geostatic compaction	60–75	Removal of free water	45–95 $W \gg W_L$
	Late	Close coagulation		35–45		30–45 $W \leqslant W_L$
Catagenesis	Early	Close coagulation and transitional	Impeded geostatic compaction	16–25	Removal of loosely bound (osmotic) water	10–18 $W \geqslant W_p$
	Middle	Transitional and phase (cementation)		4–12		3–5 $W \leqslant W_{mh}$
	Late	Phase (cementation and crystallization)	Combined development of geostatic and recrystallization compaction	2–4	Removal of tightly bound (adsorbed) water	1–2 $W \ll W_{ml}$
Metagenesis		Phase (crystallization)	Recrystallization compaction	1–2		

plastic mechanism of mutual displacement of structural elements under small shearing stresses. The sediment compaction through this mechanism proceeds at a progressively slowing rate (due to the increase in the number of contacts and their gradual strengthening at the expense of boundary film thinning with pressure growth) until the porosity values of about 25–30% are achieved.

The second stage – the stage of impeded lithogenetic compaction – begins at the end of the early catagenesis substage and continues throughout its middle substage. The compaction of clays at this stage becomes slower due to the beginning of induration and cementation and the fast growth of the number of strong transitional and phase contacts. The compaction proceeds at the expense of local rheological structural displacements, of a creep character, at the places of stress concentration, which are the largest pores and microaggregates. The minimal porosity values corresponding to the lower boundary of this stage are 4–12%.

The third stage – that of the combined development of geostatic and recrystallization compaction, beginning at the middle-to-late catagenesis substage transition and ending at the metagenesis boundary. It is characterized by the gradually increasing role of the recrystallization and authigenic neocrystallization processes in the pore space formation and by the change in the pore space

(Table 3.1. continued)

Consistence	Permeability, mD	Coefficient of compressibility, MPa^{-1}	Uniaxial compression strength, MPa	Elastic (E) and secant, (E_t) moduli MPa	Cohesion, MPa	Friction angle,°	Viscosity of undisturbed structure, Pa s
Latent-fluid	2–0.3	150–30					10^7–10^8
Fluid- and soft-plastic	10^{-1}–10^{-3}	20–0.01	0.03–0.5	$E_t = 0.1 \div 10$	0.01–0.05	5–18	10^8–10^{12}
Tight-plastic, semi-solid	10^{-3}–10^{-5}		0.6–4	$E_t = 10$–10^2 $E \approx 3 \times 10^2$	0.03–0.1	14–26	10^{12}–10^{14}
Solid			3–15	$E_t = 10^2 \div$ $\div 3 \times 10^2$ $E = 5 \times 10^2 \div$ $\div 8 \times 10^2$	0.12–0.6	20–36	10^{15}–10^{18}
	10^{-2}–10^{-3}		5–60	$E_t = 8 \times 10^2$ $= 10^3$	0.12–0.8	22–38	
	$>10^{-3}$						

character due to the development of microfractures. As a result of the combined action of clay matter recrystallization and high pressure, the porosity of rocks at this stage may become reduced to 2–4%.

The fourth stage – recrystallization of clay matter, that develops in metagenesis and is entirely associated with the internal processes of rock metamorphization. As a result of its development, the rocks acquire porosity not exceeding 1–2%.

The above scheme of the pore space transformation in clay sediments in the course of lithogenesis bears a generalized character and reveals general regularities in the development of this process. It may have its specific features in every basin depending on the composition of sediments being formed there, geodynamic regime, pressure–temperature and geochemical conditions.

3.2.   WATER CONTENT AND CONSISTENCE

In studying the subaqueous lithogenesis, clay sediments are considered as two-phase systems, i.e. as consisting of a solid and a liquid phase. Therefore, the lithogenetic change in the porosity of sediments is accompanied by the corresponding change in their water content and consistence.

3.2.1.   WATER CONTENT

Water content is determined practically in every investigation of the sediment alteration in lithogenesis. Using the water content and density data, the dry rock density and rock consistence are calculated according to Formulae (3.1) and (3.6), respectively.

There are two characteristics of water content – the *weight* (W) and the *volume* (W_v) water content, measured in per cent. The first index is used more often. It characterizes the ratio of the weight content of water mass to the mass of dried rock. The second index shows the volume of water in a volume unit of a rock with undisturbed structure and is commonly used to estimate the degree of pore filling with water. There is a simple interrelation between the two indices:

$$W_v = W \times \rho_d \qquad\qquad (3.5)$$

where ρ_d is matrix (dry rock) density.

In studying the lithogenesis of sediments, it is most interesting to determine not just the total mass of water in the clay, but rather the contents of its individual categories – the free (gravity) and the physically bound water, including the adsorptionally bound water of the external surfaces and interlayer intervals of clay minerals. It is the energy and structural heterogeneity of water contained in a clay deposit that makes dehydration of rocks one of the most important processes affecting the development of lithogenesis in a cardinal way. The changes in practically all the properties of clays are also connected with this factor. So, due regard for it is a matter of principle in considering the clay rocks as seals for oil and gas fields.

Direct methods for determining the different categories of water in clays are absent or require much labor. Therefore, indirect indices giving approximate values of different water categories are often used in practice. There are such indices of water content used on a large scale, as the swelling limit (W_s), plastic limit (W_p) and liquid limit (W_L), maximal hygroscopic water content (W_{mh}), which are determined following the standardized techniques [22].

The swelling limit is believed to characterize the mass of water that clay is able to retain due to the adsorption, capillary, and osmotic forces. Therefore, this water content is identified with the total amount of physically bound water in a rock [3]. Within the water content range of W_p to W_L, a diffuse ionic layer of clay minerals forms. So the difference $W_p - W_L$ may be a rough index of the amount of osmotically and capillary bound water in a clay. The maximal hygroscopic water content (W_{mh}) shows the total amount of water which a clay may adsorb from air at $\rho/\rho_m \cong 0.9$, and thus may serve as a characteristic of the tightly bound (adsorptionally bound) water in a clay [3].

The forming clay sediments contain all the categories of water, beginning with the adsorptionally bound and ending by free water. The total water content in sediments may be as high as 150–350% by weight and 80–85% by volume. The bulk of water mass is free (gravity) water contained in large cellular pores of the sediment. The content of physically bound water at the initial substage of diagenesis usually does not exceed 15–25% of the total water content in sediments and may attain 30–40% only in the sediments of montmorillonite composition.

At the early and late diagenesis substages, under the conditions of free geostatic compaction of sediment, the free water is being squeezed out. The squeezed-out water moves, as a rule, up the section into the near-bottom part of water basin or into the coarser overlying sedimentary layers which serve as draining horizons. The shallow-water sediments having coarser composition and often containing interlayers of sand and silt material are dehydrated much more quickly. In the homogeneous sequences of clay sediments deposited in the middle and deep-water shelf parts of inner and marginal seas, as well as on the continental slopes of oceans, the rate of water squeezing-out slows down because of low filtration properties of these sediments. The presence of decomposed vegetable organic matter also contributes to the slowing of this process. The squeezing-out is particularly slow in the deep oceanic sediments having small-honeycomb structure and considerable contents of swelling minerals.

In the process of diagenesis, a sediment loses up to 70–80% of all the water stored in it. Its water content becomes equal to, or slightly lower than, that of the liquid limit, making up most often 30–40%. The free water proves to be squeezed out. It is mainly physically bound water that remains in the sediment. A certain volume of free water may only be preserved in the largest pore voids ranking in size as small micropores. The filling of these pores with free water proceeds largely at the expense of loosely bound (mainly osmotic) water which is squeezed out into these pores from the closing fine micropores and ultramicropores.

In the subzone of early catagenesis, the remaining free water is removed and the squeezing-out of osmotically bound water proceeds. The presence of this category of water in the clay is mainly caused by the existence of a diffuse layer of ions around clay particles. The diffuse layer thickness attains several hundredth and even tenth fractions of μm. It increases with the growth of temperature, and the diffuse layer boundary becomes more and more blurred. The osmotically bound water forms a kind of «cuffs» inside of pores, significantly reducing their «live» cross section, i.e. that part of pores in which free water can move without difficulty. Ultramicropores (<0.1 μm) and probably part of fine micropores (0.1–1 μm) prove to be completely obstructed by osmotic water and are thus excluded from the open porosity. With the beginning of closing of small and fine micropores and compaction of microaggregates at the early catagenesis substage, the diffuse layer ions together with water molecules hydrating them begin to be squeezed out beyond the clay layer. This process leads, on the one hand, to the increase in concentration of the pore solution being squeezed out and, on the other, to the intensification of diffusional transfer of poorly hydrated cations (including K^+ and Si^{4+}) towards the clay sequence. The conditions are being created for the beginning of clay cementation at the expense of pore solution oversaturation in silicon and for clay mineral hydromicatization at the expense of potassium surplus.

The process of the squeezing-out of loosely bound water may continue until the isothermal boundary of 65–70 °C is reached, which corresponds in most cases to the middle of the middle catagenesis subzone. The water content of rock at this boundary becomes lower than that of the plastic limit and approaches the maximal hygroscopic water content. Depending on the grain size distribution and mineral composition, its value may be 5–8%. Thus, from the beginning of the early substage of catagenesis and up to the middle of its middle substage, all the loosely bound (osmotic) water, whose total amount is about 12–20% of the initial water reserve in the sediment, is removed from the rock.

The squeezing of osmotic water out of the rock differs in its mechanism from the free water removal. The latter proceeds solely under the action of hydrostatic (pore) pressure arising in the rock owing to the compaction of its matrix. The removal of osmotic water, due to its higher viscosity, occurs only under high effective stresses and, most important, under the conditions of the counteracting osmotic mass-transfer process. So, dehydration of clay rocks at the early substage and up to the middle of the middle catagenesis substage is not just a mechanical, but also a physicochemical (diffusion) process.

One may suppose that it is just under these conditions that the highest screening properties of clay rocks form, resulting in the rise of anomalously high formation pressures (overpressures). This conclusion is based on the following considerations. Firstly, within the discussed interval of lithogenesis, the clay structure becomes most homogeneous owing to the destruction in it of all stress concentrators represented by the largest pores and microaggregates. This results in the increase in the clay pore space of the number of ultramicropores and fine micropores blocked to a considerable degree by adsorptionally bound water and not participating in the filtration transfer of water. Besides, at the contact of such a clay with oil and gas, the screening role of capillary pressure, which may attain 15 MPa or more, sharply increases. Secondly, the mechanism of rock formation within this interval of lithogenesis, as has been said above, keeps its plasticized (creep) character and is not accompanied by the break of rock continuity, i.e. the conditions for microfracture formation are absent. And thirdly at last, the squeezing-out of water is impeded because of the osmotic water transfer in the opposite direction and the enhanced viscosity of physically bound water.

According to O.A. Martirosova and V.V. Bakhtin [139], the presence of overpressure hinders the clay compaction, which may result in the relatively high clay porosity values at depths where overpressures are observed or at slightly greater depths. I.D. Zkhus and V.V. Bakhtin [103] came to similar conclusions.

After the isothermal boundary of 65–70 °C is crossed, a new stage of clay dehydration begins. It is determined by the beginning removal of tightly bound (adsorptionally bound) water. The mechanism of this process is of thermodynamic nature and is not connected with the acting pressure. It is based on the thermal activation of water molecules retained by the adsorption forces on the surface of minerals. With the growth of temperature, the frequency of translational jumps of water molecules increases, and thus the time of their presence near the surface decreases. After the temperature of 65–70 °C is reached, the number of translational displacements of adsorbed water molecules becomes equal to the number of these events in free water, and the adsorptionally bound water passes to free state.

The passage of tightly bound water to free state affects significantly the entire course of lithogenesis, accounting for a number of effects accompanying this process. An important role belongs to the enhanced dissolving power of the forming free water owing to the protonization (dissociation) of water molecules caused by the strong polarizing action of counter-ions of the electrical double layer [98]. Dissolution of salts and some silicates gives rise to an increase in the pore solution concentration and to the activation of clay cementation process at the expense of crystallization of new mineral phases. It is not excluded that the long-term predominance of dissolution and substance removal over the formation of new minerals may lead to the decompaction of rocks and weakening of their structural bonds. In this case, the formation of anomalous horizons in a clay sequence is possible, as described by Z.A. Krivosheeva [93].

In general, however, the transformation of adsorptionally bound water into free one leads to an increase in the cementation and strength of clays and, what is most important, determines their brittle (rather than plasticized) destruction and origination of microfractures at the places of stress concentration. The rise of brittle destruction in rocks is much facilitated by the reduction of Rebinder's effect [149] which causes a decrease in the strength of solids and their adsorptional plasticizing in deformation. The dehydration of adsorbed water molecules reduces the action of this effect and contributes to the strength increase similar to the strengthening of clay during its air drying.

It is of great importance that the «live» cross section of micropores increases at the expense of boundary films of bound water disappearing from them. If one takes into consideration that viscosity of free water at the temperature of 65–70 °C is several times as low as its viscosity at normal temperature, then, in spite of the relatively low porosity, the pore permeability of clay rocks increases with the removal of adsorptionally bound water. This is also assisted by the process of microfracture formation. All this indicates that beginning with the middle of the middle catagenesis subzone, deterioration of the screening properties of clay seals may start.

The process of dehydration of adsorptionally bound water develops, due to its energy heterogeneity, up to an isotherm of 110–120 °C. This means that it may also continue in the late catagenesis subzone. Water molecules adsorbed at the mineral edges and at the cations of swelling interlayer intervals have the highest energy. It is not unlikely that the dehydration of interlayer cations, along with the processes of isomorphism, favors the hydromicatization of montmorillonite and mixed-layer minerals, which, according to the data of a number of authors [79], develops progressively beginning with the middle catagenesis substage.

At the late catagenesis substage, the screening properties of clay rocks continue to deteriorate. This effect becomes more and more pronounced as the argillization of rocks and their structural rearrangement develop, with large platy polycrystalline overgrowths being formed. Concurrently, the heterogeneity of structure and structural bonds increases, which promotes further development of microfractures along the boundaries of individual polycrystalline overgrowths.

Thus, by the end of catagenesis, the dehydration of rocks is practically completed. The residual water content at the metagenesis boundary does not exceed 1–2%. During the second half of the middle substage and the entire late catagenesis substage, the last portions of water that were most tightly retained in the rock are removed. The total amount of tightly bound water removed from the rock at these lithogenesis substages makes up 4–5% of the original water reserve in sediments.

Investigation of the whole process of water removal from clay sediments in the course of lithogenesis makes it possible to differentiate in this process three stages associated with different energy states of water and mechanisms of its dehydration (see Table 3.1).

The first stage – removal of free water under the action of increasing geostatic pressure continuing throughout the diagenesis. At this stage, practically all free (gravity) water enclosed in large and small micropores of sediments is squeezed out. By the moment of its termination the sediment loses about 70–80% of originally stored water. The water retained in the sediment is predominantly in a physically bound state. The water content of sediment is close to that of the liquid limit or slightly lower and makes up 30–45%.

The second stage – removal of loosely bound (osmotic) water. Dehydration of rocks at this stage is of mechanical and physicochemical nature and proceeds under the influence of the opposing processes: the squeezing out of water under the action of growing geostatic pressure and the osmotic mass transfer directed towards clay units. The second stage of dehydration continues to a depth with an isotherm of 65–70 °C that approximately corresponds the middle of the middle catagenesis subzone. At these depths the osmotically bound water is completely removed. The water content of rocks becomes close to the maximal hygroscopic water content and has values of 5–8%. The total amount of removed loosely bound water is 12–20% of the initial water content of the sediment.

The third stage – removal of adsorptionally bound water by means of its transformation into free water and squeezing of the latter out of the rock. The basis of this dehydration stage is a thermodynamic process: an increase in the number of translational displacements of adsorbed water molecules with the growth of temperature and their passing into free state. The process of adsorptionally bound-to-free water transformation activates the mineral dissolution and neocrystallization, reduces the Rebinder's effect, favors the development of microfractures. The third stage of dehydration continues during the second half of the middle substage and throughout the late substage of catagenesis. By the end of catagenesis, practically complete dehydration of rocks occurs. The residual water content does not exceed 1–2%. The total amount of tightly bound water removed at this stage makes up 4–5% of the initial water content of the sediment.

3.2.2. CONSISTENCE

Consistence determines the character of clay deformation and destruction. It depends on the water content and structural bonds.

For a quantitative estimation of consistence, the natural water content of the clay (W), the plastic limit (W_p) and the liquid limit (W_L) determined according to Atterberger are used. Calculation of the *consistence index* (I_c) is done using the following formula:

$$I_c = \frac{W - W_p}{W_L - W_p} \tag{3.6}$$

Depending on the value of I_c, the following clay consistencies are differentiated: fluid ($I_c > 1$), fluid-plastic ($0.75 < I_c < 1$), soft-plastic ($0.5 < I_c < 0.75$), tight-plastic ($0.25 < I_c < 0.5$), semi-solid ($0 < I_c < 0.25$), and solid ($I_c < 0$) [3].

It is expedient to judge on the consistence of clays occurring at different stages of lithogenesis not only from the I_c value, but also from the predominant type of contacts. With a change in the latter, the consistence of clay deposits and the character of their deformation also change. Thus, sediments with pre-dominance of distant coagulation contacts show fluid or latent-fluid consis-tence, while those with predominance of close coagulation contacts show plastic consistence. Clays with coagulation and transitional (or phase) contacts have semi-solid consistence. At last, the predominance of transitional and phase contacts results in the solid consistence of rocks. Taking into account the above information, one may estimate the consistence of clay rocks at different stages of lithogenesis (Table 3.1).

In the subzone of early diagenesis, sediments have fluid consistence in the disturbed condition and latent-fluid consistence in the undisturbed condition of structure, as the close and distant coagulation contacts are prevalent in them and their water content exceeds the liquid limit.

In the subzone of late diagenesis, water content of clays approaches the liq-uid limit: in the upper part of the subzone it is higher than W_L, while at its base it becomes slightly less than, or equal to W_L. The close coagulation contacts become predominant. In the undisturbed condition of structure, sediments show strongly pronounced plastic (from fluid-plastic to soft-plastic) consistence.

With the passage to the subzone of early catagenesis, the coagulation con-tacts become gradually replaced at first by transitional and then by phase (cementation) contacts. Water content of deposits at the base of the subzone approaches W_p or remains slightly higher than it. The clay consistence also undergoes corresponding changes: from tight-plastic in the upper part it passes to semi-solid at the base of the subzone.

Beginning with the subzone of middle catagenesis, the consistence of rocks becomes solid. This is connected with the predominance in them of strong, at first transitional and cementation and then cementation and crystallization contacts. The water content acquires values lower than W_{mh}.

The above-described character of the consistence change has been obtained for clay rocks taken from various depths and tested under normal pressure-temperature conditions. As the behavior of rocks under deformation is influenced by temperature and pressure, their real consistence under natural conditions of occurrence may differ from the described above. Therefore, data obtained under laboratory conditions should be considered as an especially qualitative characteristic that may be used for a preliminary classification of clay rocks according to degree of their lithification.

3.3.   PERMEABILITY

Permeability is understood as an ability of a porous medium (rock) to let fluids and gases pass through it [150]. Permeability characterizes the filtration properties of rock, which are determined by the *coefficient of filtration K* and *permeability index* K_p. They are interconnected as follows:

$$K = K_p(\gamma_w/\mu)$$

(3.7)

where γ_w and μ are density and viscosity of the filtering liquid. The coefficient of filtration is measured in m/day.

Thus, the coefficient of filtration is an integral index reflecting the properties of both the rock (through the permeability index K_p) and the filtering fluid (through its viscosity and density).

The permeability index is determined using the following formula [150]:

$$K_p = d_e^2 Sl_{(n,e)}$$

(3.8)

where d_e is the effective diameter of particles composing the porous rock, and Sl is Slichter's number, a dimensionless value determined by the porosity value and pore space structure. The permeability index is measured in Darcy (D) or in μm^2 (1D = 1 μm^2). For fresh water, $K_p = 1$ μm^2 corresponds to $K = 0.86$ m/day.

The process of filtration in clay rocks is exceptionally complicated. This is caused in the first place by the fact that clay represents a fine grained, highly porous system with extremely large specific surface. It is composed predominantly of clay mineral particles having anisometric shape, micron and submicron sizes, characterized by specific crystallochemical structure and showing particular behavior when interacting with water. An important role in the formation of filtration properties is attributed to bound water that fills the pore space.

Porosity n is the main parameter determining the value of index Sl and permeability as a whole. The index Sl depends also on the pore space structure characterized by a dimensionless coefficient of tortuosity ε. The coefficient of tortuosity is understood as the ratio of the least distance between two points in the direction of filtration to the length of the tortuous path of fluid flow through pore channels.

Various authors used the notion meant by the coefficient of tortuosity under different names: «hydrodynamic coefficient of tortuosity» [142], «lithologic coefficient» [151], «dividing coefficient» [152].

Although Formula (3.8) contains only the parameters of a porous medium, it is known from practice that the permeability index of rock changes significantly depending on the nature of a fluid, its chemical composition, thermodynamic conditions, and hydrodynamic pressure. Such changes in the rock permeability are accounted for by the processes of physicochemical interaction of the filtering solution with the mineral matrix of the rock, change in the pore fluid properties under the influence of temperature, mobilization of the near-wall layers of fluid at higher gradients, reorientation of mineral particles, and redistribution of pore sizes.

Permeability of one and the same rock to a gas is usually significantly (in some cases, several hundred and thousand times) higher than to a liquid. This is caused by the lesser effect of interaction between gas and mineral matrix.

3.3.1. FACTORS AFFECTING THE PERMEABILITY

One of the main factors affecting the permeability of clays is the presence of bound water in the pore space. Its influence is due to the fact that the thickness of a bound water layer determines the volume of free pore space (active porosity), just on which the rock permeability is principally dependent.

Filtration of water through clay rocks proceeds along the largest pore channels (interconnected pores) not entirely filled with bound water. In the pore channels of lesser sizes, when a significant pressure gradient is present, part of loosely bound water is squeezed out, and so these pores can also participate in water filtration. The smallest pores (<0.1 μm in diameter) are entirely occupied with tightly bound water. Therefore, even at high pressure gradients, they do not practically filter water.

Earlier, the structure of pore space in clay rocks has been described and the peculiarities of its transformation in lithogenesis have been considered. So, in the present section, the main attention will be given to the consideration of factors determining the formation of active pore space in clays. They include: mineral composition, salinity and chemical composition of the filtering fluid, temperature and pressure gradient, geostatic pressure.

Mineral composition.   Investigations by I.A. Briling [153], V.M. Gol'dberg and N.P. Skvortsov [150] have shown that permeability highly depends on the mineral composition of clays and the composition of exchange cations. According to I.A. Briling (Table 3.2), the *Na*-form of montmorillonite clay at water content of W_p has 10 times lower permeability than *Na*-kaolinite. The permeability of hydromica has an intermediate value. A change of Na^+ for Ca^{2+} in the exchange ion composition makes the permeability 5–10 times as high.

Table 3.2. Permeability of monomineral clays.

Mineral composition of clays	Kp, mD
Na-kaolinite	2.3×10^{-2}
Ca-kaolinite	1.1×10^{-1}
Na-montmorillonite	2.3×10^{-3}
Ca-montmorillonite	1.1×10^{-2}
Hydromica	6.2×10^{-3}

Such a dependence is explained by the peculiarities of clay particle size distribution, crystallochemical structure, and hydration of clay minerals. Thus, montmorillonite, having an expanding crystal lattice, small size of particles, large specific surface and exchange capacity, contains the volume of bound water exceeding many times that of kaolinite. The latter, in its turn, has a rigid crystal lattice and large microcrystalls, this causing its relatively small specific surface and exchange capacity. Hence, the kaolinite clays must have more «open» pore space, which leads to higher permeability values. A similar effect occurs with the change of Na^+ for Ca^{2+} in the exchange complex, when the coagulation of clay particles and compaction of diffuse layers are observed, this leading to an increase in the active pore space for the filtration of free water. The effect intensifies with a change of the exchange complex for the higher-valence cations with heavier atomic weights.

And on the contrary, saturation of clay with univalent cations leads to an increase in the rupture of clay aggregates, decrease in the diameter of pore channels, and reduction of permeability. This effect is the greater, the less the ionic radius of a cation.

Salinity and chemical composition of filtering fluid. This influence is manifested through change in the composition of absorbed ions in a clay deposit and the processes of coagulation – dispersion of clay particles.

As the salinity of the filtering fluid grows, the diffuse layers around clay particles become intensely suppressed, which leads to a decrease in the volume of bound water. Concurrently, the active pore space increases resulting in the increase in clay permeability. When fresh water or a solution of lesser salinity than that of the pore fluid is filtering through clays, the opposite effect takes place, accompanied by the growth of diffuse layer thicknesses and increase in the amount of bound water. This results in a decrease in the active pore space volume and reduction of clay permeability.

In solution filtering through clays, two types of solution – rock interaction can be discussed. The first one – when the filtering solutions contain the same cations which are present in the exchange complex. In this case, the cation exchange is practically absent, and the exchange complex composition does not change. Therefore, the change in the diffuse layer thickness will be mainly determined by the pore solution concentration, or, more exactly, by the difference of concentrations of the filtering fluid and the pore solution. The second type – when the filtering solution contains cations that are absent in the exchange complex. In this case, the change in permeability will be of a more complicated nature. On the one hand, the permeability change will be determined by the kind of a new cation entering into the exchange complex, on the other – by the concentration of the filtering solution. Different versions of such a double effect are possible here.

The highest dispersion degree and the maximal development of diffuse layers are characteristic of the clays, in whose exchange complex sodium ions predominate. These clays must have minimal permeability and the widest range of its variation when they filter solutions of different concentrations. The least dispersion degree and the minimal development of diffuse layers are characteristic of clays with predominance of multivalent cations (Ca^{2+}, Al^{3+}, Fe^{3+}) in the exchange complex. These clays are expected to have maximal permeability.

Thus, when a solution is filtering through a clay, if Na^+ cations appear in the exchange complex as a result of cation exchange, a decrease in permeability is to be expected, and if these are Ca^{2+}, Al^{3+}, or Fe^{3+} cations – an increase in permeability will follow. Besides, the diffuse layer thickness will also become suppressed with the growth of pore solution concentration. So the permeability increase will be maximal with the growth of concentration of a solution containing multivalent cation salts.

Dependence of permeability on the salinity and composition of filtering solution was analyzed by V.M. Gol'dberg and N.P. Skvortsov [150]. The results of their investigations have shown that permeability of Na- and Ca-montmorillonite and kaolinite grows with the growth of concentration (C) of NaCl and CaCl solutions from 0 to 150 g/l. The maximal increase in permeability has been shown by Na-montmorillonite clays: when a distilled water was changed for $CaCl_2$ ($C = 55$ g/l), the K_p value increased 13 times (from 0.8×10^{-4} to 10.4×10^{-4} mD), and when a distilled water was changed for a NaCl solution of the same concentration, it increased 4.2 times (from 0.7×10^{-4} to 2.9×10^{-4} mD). In Ca-montmorillonites under the same conditions, permeability increased 2 times for NaCl solution (from 2.1×10^{-4} to 4.2×10^{-4} mD) and 2.7 times for CaCl (from 1.3×10^{-4} to 3.5×10^{-4} mD).

For a natural kaolinite clay, within a salinity range of 0 to 150 g/l, the permeability increases up to 2.7 times (from 0.9 to 2.4×10^{-3} mD) with a NaCl solution being filtered and up to 2.3 times (from 1.0 to 2.3×10^{-3} mD) with $CaCO_2$ solutions.

The influence of chemical composition of the filtering solution has been studied for salt concentrations growing from 0 to 150 g/l. It has been shown that its minimal effect on the permeability increase observed in natural clays of kaolonite–montmorillonite composition is as follows: in filtering a NaCl solution – 1.23 times (from 0.07×10^{-3} to 0.086×10^{-3} mD); in filtering a KCl solution – 1.45 times; $CaCl_2$ – 1.75 times. A very sharp increase in K_p is observed in filtering a $FeCl_3$ solution –12.5 times. According to the degree of their influence on the intensity of the permeability index growth (with salinity being the same), the chloride salts range as follows:

$$FeCl_3 > CaCl_2 > KCl > NaCl \qquad (3.9)$$

Taking into account that solutions of these salts have the same anion composition, the cation components of this series may be written in the following sequence:

$$Fe^{3+} > Ca^{2+} > K^+ > Na^+ \qquad (3.10)$$

This series coincides completely with the similar series of cation activity and the energy of their absorption by clays. Taking into consideration that, as the cation valence grows, the content of tightly bound water in clays increases along with a concurrent significant decrease in the content of loosely bound water [3], one can explain the change in the permeability of clays depending on the composition of filtering solution by the effect produced by exchange cations on the thickness of a diffuse layer around particles, the volume of active porosity being dependent, in its turn, on this thickness.

Temperature and pressure gradient. The influence of temperature on clay rock permeability manifests itself in the change of free and bound water viscosity. With the growth of temperature, due to the increase in the energy of thermal movement, the viscosity of loosely bound water and free water decreases, which favors the increase of clay permeability. This effect is most pronounced in montmorillonite clays containing a large volume of bound water. V.M. Gol'dberg and N.P. Skvortsov [150] show in their work that, with temperature growth from $20\,°C$ to 80–$90\,°C$, the permeability to fresh water in montmorillonite clays increases 12.5 times (from 0.057×10^{-3} to 0.72×10^{-3} mD) and to filtering chloride solutions – 35 times (from 0.316×10^{-3} to 11.07×10^{-3} mD). The permeability of clays of kaolinite composition increases no more than 2–3 times (from 0.88×10^{-3} to 2.78×10^{-3} mD), and of hydromica clays – 5–6 times (from 0.39×10^{-3} to 2.24×10^{-3} mD).

The most intense permeability increase is observed at about $65\,°C$. This is accounted for by the transition of adsorptionally bound water into free state and an increase in the «live» cross section of pores. At these temperatures, the initial gradient of filtration is practically absent and filtration obeys Darcy's law.

The influence of pressure gradient on clay permeability is determined by the viscous-plastic properties of bound water and heterogeneity of its connection with the mineral surface of clay particle. Under the initial filtration gradient, the viscous-plastic resistance to shearing of the most external layers of a bound water film is overcome and this water begins to participate in the filtration process. As the pressure gradient grows, larger and larger volume of bound water becomes mobilized. This leads to the growth of the effective cross section of pore channels, through which the fluid moves. With a decrease in the pressure gradient, the effective cross section tends to compress elastically. Thus, a bound water film in the cross section of a pore channel may be considered as an elastic membrane. The opening of pore channels increases with the growth of pressure gradient owing to the mobilization of more and more bound water and decrease in thickness of the remaining film of immovable bound water. And on the contrary, with a decrease in pressure gradient, the opening of pores to water movement decreases due to the thickening of the immovable bound water film.

Geostatic pressure. The action of geostatic pressure in clays leads to a decrease in the pore space volume as a result of clay particles and their microaggregates turning normally to the applied load and becoming more closely packed. When there is upward or downward filtration, this process is opposed by the hydrostatic pressure of water tending to turn the particles in the flow direction, i.e. along the compacting force. Such a reorientation of particles is most easily realized in loose water-saturated sediments and leads to some increase in filtration. In compacted clay rocks, the turn of structural elements is impeded. The growth of pressure in such clays may result in the breakthrough and destruction of the rock. The break of clay seals occurs under the water head of several thousand meters and has been called the «breakdown pressure».

Data available on the determination of clay rock permeability at depth show a definite decrease in permeability with the growth of geostatic pressure. The results of investigations reported by V.M. Gol'dberg and N.P. Skvortsov [150] evidence that an increase in geostatic pressure from 3 to 20 MPa causes a 17.5 time decrease in clay permeability (from 15×10^{-4} to 0.85×10^{-4} mD). With a pressure change from 3 to 6 MPa, the permeability of a montmorillonite clay decreased 2.1 times, of argillite – 2.3 times, and of a kaolinite clay (with pressure growth from 1 to 2 MPa) – 1.5 times. The most intense decrease in clay rock permeability occurs at the initial stage of pressure growth.

When there is a concurrent increase in geostatic pressure and temperature, permeability also shows a tendency of reduction, which is caused by the intense compaction of rocks and decrease in their porosity [153]. A decrease in free water viscosity is not able to compensate the quick reduction of rock porosity.

3.3.2. CHANGE IN CLAY ROCK PERMEABILITY IN LITHOGENESIS

The formation of porosity and permeability of clay rocks is closely associated with the process of lithogenesis, being one of its characteristic features. In solving this problem, one should proceed from the notions about the pore space structure and presence of different categories of bound water. Unfortunately, no data can be found in scientific literature describing a change in clay rock permeability in lithogenesis for a single sequence within a large depth range. Therefore, in the present section, the main trends in the change of clay rock permeability in lithogenesis will be described based on theoretical considerations and illustrated, when possible, by individual experimental permeability values taken from various sources.

The processes occurring at the substage of early diagenesis are intense compaction and dewatering of sediments whose permeability at shallow (up to 10 m) depths may be 0.3–2 mD [154, 155].

With clay sediment compaction at the substage of late diagenesis, the structure of pore space undergoes considerable changes: porosity decreases, large intermicroaggregate micropores (10–100 μm in size) practically disappear and small intermicroaggregate micropores (1–10 μm in size) decrease in content, reorientation proceeds, with structural elements increasing their degree of orientation in the direction perpendicular to the load. The bulk of free water is squeezed out, with the rest of it being preserved only in the largest pores. The geostatic pressure leads to significant reduction of active pore space and considerable decrease in permeability, in spite of temperature growth and some decrease in water viscosity. According to V.M. Gol'dberg and N.P. Skvortsov [150], clay deposits at this substage of development decrease their permeability to 0.29×10^{-3} mD.

In catagenesis, the permeability is considerably influenced by thermodynamic factors (pressure and temperature). As a result, porosity and permeability continue to decrease in the subzone of early catagenesis. It is necessary to note a wide scattering of permeability values given by different authors for the clay seals occurring at this substage of lithogenesis. Thus, A.A. Khanin [140] gives values of 5×10^{-3}–5×10^{-4} mD for clay seals of the Amu-Darya and Central-Turkmenian oil- and gas-bearing regions; for the Western Pre-Caucasus region and Fergana, the scattering of values is 10^{-2} to 10^{-5} mD [138]; for clay seals of the Okhotsk oil- and gas-bearing province it is 10^{-2}–10^{-3} mD [109]. This is connected with the fact that at the early substage and beginning of the middle substage of catagenesis, the seals with minimal permeability values form, whose screening properties are determined, as has been noted earlier, by the presence of fine and small micropores filled to a considerable extent with bound water. The pore space structure, in its turn, depends on the mineral composition, exchange cation composition, pore solution salinity, pressure gradient, geostatic pressure, and so on. All these factors, no doubt, manifest themselves in different ways in various regions, and most unexpected combinations of them are possible. This may be the reason for the great scattering of clay seal permeability values. A good example of the aforesaid is the permeability of seals in Fergana, where clays of montmorillonite–mixed-layer composition had permeability of 10^{-5} mD, while the hydromica-kaolinite varieties characterized by lower total porosity but having a greater active porosity volume showed higher permeabilities of 10^{-2}–10^{-3} mD [138].

At the substages of middle and late catagenesis, with the passage over the isothermal boundary of 65–70°, some increase in permeability may be expected. Such a behavior of clay rocks may be related to the transformation of adsorptionally bound water into highly aggressive free water. This is accompanied by an increase in the «live» pore cross section and dissolution of mineral substance, which creates favorable conditions for the formation of anomalous horizons and increase in their permeability. The increase in permeability in the zone of anomalous horizons results in the loss by clay seals of their screening properties.

In the process of the following catagenetic transformation of clay rocks, a tendency of further porosity reduction is observed. At the same time, at the end of the middle – beginning of the late catagenesis substages, processes of clay rock argillization begin, leading to the enlargement of solid structural elements and formation of polycrystalline overgrowths of a platy shape. This is accompanied by the growth of stress concentrators, origination and the following development of microfractures. As a result, significant changes occur in the pore space structure of clay rocks, where, along with the preserved intermicroaggregate pores of a fissure-like shape, new paths of filtration through microfractures begin to form. Such a filtration mechanism often brings to some increase in the permeability of clay seals at great depths. V.V. Bakhtin et al. [156] report data on the argillites of Carboniferous age recovered from a depth of 3083 m at a temperature of 100 °C within the Solokhovskaya area in the Ukraine. Petrographic studies of thin sections of these rocks have revealed the presence of multiple microfractures with openings up to 10 μm wide. These rocks with porosity of 3.8% had permeability of 6×10^{-2} mD.

The combined effect of high pressures and temperatures in late catagenesis results in the practically complete dehydration of clay rocks. This fact favors the development of phase contacts between solid structural elements and completion of the process of clay rock argillization. At this stage of lithogenesis, intense formation of microfractures proceeds in argillites and, the porosity being very low (2–4%), the permeability is mainly provided by microfractures. The seals for hydrocarbon fields represented by argillites often lose their screening properties, as is evidenced by data of G.E. Kuz'menkova and A.A. Fomin [157]. These authors note that the samples of fractured argillites recovered within the Levkinskaya area of the Pre-Caucasus region from a depth of 4900 m at a temperature of 140 °C, had permeability about 1×10^{-3} mD, with porosity being 3.4%.

At the metagenesis stage, under the action of high pressure and temperature, the main processes are the recrystallization of clay matter, closing and healing of microfractures accompanied by a decrease in porosity sometimes to as low values as 1–2%. However, because of the high brittleness of clay rocks and their rather intense tectonic fracturing, filtration in these beds occurs most often according to the scheme of a heterogeneous system consisting of low permeable clay blocks and relatively highly permeable zones at the block contacts.

Unfortunately, the authors have no available data on the permeability values in metamorphosed clay rocks. It may be supposed, however, that these rocks form the seals of low classes with permeability higher than 10^{-2} mD.

In conclusion, it should be noted that permeability is the principal, but very complex and many-factor parameter determining the screening properties of clay seals for hydrocarbon pools. With the existing general trend of decreasing permeability in the process of lithogenetic alterations, there are a number of reasons that break this regularity. The main of them are the anomalous properties of bound water at high pressures and temperatures and the presence of microfracturing.

3.4. PHYSICAL-MECHANICAL PROPERTIES

As has been already mentioned above, destruction and deformation of clay rocks occurs along the contacts of structural elements, as the strength of contacts is much less than that of the minerals composing the rock. So, the character of structural bonds determining the energy type of the rock structure is critical in estimating the strength and deformation properties of clay rocks. Proceeding from this, some peculiarities of the physical-mechanical properties of clay rocks depending on their structural features will be discussed below.

3.4.1. COMPRESSIBILITY

Compressibility of clays is characterized by the value of their porosity change related to the load range under which the deposit has been compacted.

The estimation of compressibility may be performed following two schemes. The first scheme, that has been termed the uniaxial compression, consists in the uniaxial compression of a clay in a hard case excluding the lateral expansion of the sample. Compressibility in this case is estimated by the *coefficient of uniaxial compressibility* (a_v), which is found following the formula:

$$a_v = \Delta e / \Delta \sigma \qquad (3.11)$$

where $\Delta \sigma$ is the load range for which the calculation is being done, Δe is the corresponding change in the void ratio.

The second scheme is more perfect and provides the testing of rocks under triaxial compression. In this case, both the vertical and the lateral deformation of sample can be measured. When an adequate equipment is available, testing may be conducted at different temperatures, which allows modeling the stressed state of rocks corresponding to the conditions of their lithogenesis. Compressibility is estimated in this case using the *bulk modulus* (K) calculated following the formula:

$$K = \Delta \sigma_0 / \varepsilon_v \qquad (3.12)$$

where σ_0 = confining pressure value, ε_v = relative volume strain.

In spite of the advantages of the second scheme, the uniaxial compression is most commonly used in compressibility testing. This is explained by the simplicity of the technique and equipment required for such tests.

An analysis of available data on the uniaxial compressibility evidences that compressibility of no other rock varies within so wide a range as that of clays. Depending on the character of structural bonds, the coefficient of uniaxial compressibility for clay deposits may vary from 1–150 MPa^{-1} to the ten-thousandth MPa^{-1}. The highest compressibility is characteristic of muds having a high-porosity structure with a considerable amount of distant coagulation contacts. As the structural bonds become stronger, the compressibility value of clays decreases; many indurated clays, claystones, and siltstones may be considered as practically incompressible rocks. However, such an estimation is somewhat subjective, as the compression tests are short-term and do not allow estimating the compressibility of cemented clays at the expense of rheological processes.

The uniaxial compression tests are conducted in special compression apparatus. Most of the serially produced compression apparatus allow the compression of rock under a vertical load of no more than 3–5 MPa when the lateral expansion of rock is absent. Such pressures model a natural load existing only in the diagenesis zone. Therefore, for the compression tests of clay rocks having deeper occurrence, special constructions of apparatus with powerful presses are used. A number of researchers used apparatus in which the uniaxial compression of samples under loads as high as 200–300 MPa had been reached [94, 95].

A large amount of data is available in the literature on the compression tests of clay deposits from the diagenesis zone, as well as of artificially produced clay

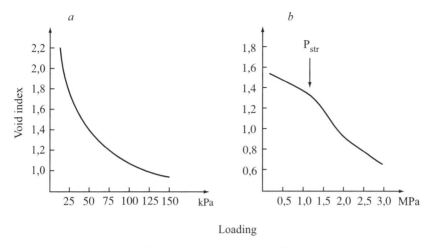

Fig. 3.6. A plot of the uniaxial compressibility of clay sediments from the early (*a*) and late (*b*) diagenesis subzones.

pastes and sediments [51, 96, 127, 158, 159]. All of them evidence that the value and character of the compressibility curve for such sediments are determined by the peculiar behavior of coagulation contacts. As one can see from Figure 3.6, *a*, the uniaxial compression plot for muds taken from the early diagenesis subzone has the form of a steep curve gradually flattening with the growth of loading. The compressibility attenuation is connected with the increase in the number of contacts and gradual transformation of distant coagulation contacts into close ones with an increase in the sediment density. The value of the coefficient of uniaxial compressibility is $150 \, \text{MPa}^{-1}$ at the initial compaction loading and decreases to $30 \, \text{MPa}^{-1}$ at the final loading (0.15 MPa). It is important to note that the graph shows no bend evidencing for the presence of structural strength.

The compressibility of plastic clay sediments from the late diagenesis subzone (Fig. 3.6, *b*) decreases noticeably compared to the muds from the overlying subzone, which is related to the predominance of close coagulation contacts in them. A characteristic feature of their compressibility is the presence on the deformation curve of a bend point (P_{str}) corresponding to the structural strength of the sediment. The structural strength value under a normal cycle of subaqueous lithogenesis is determined by the maximum geostatic loading that has affected the sediment. In geotechnics, such clays are commonly considered as «normally consolidated». In the uniaxial compression testing of a sample, the initial portion of the compressibility curve corresponds to the repeated compression on the unloading branch and has a relatively small slope. After the loading corresponding to the P_{str} point is exceeded, the compressibility curve slope slightly increases. At $P > P_{str}$, the sample shows the compressibility of sediment, which is to be expected with the further growth of geostatic compaction.

As it follows from the aforesaid, the compressibility of plastic clay sediments from the late diagenesis subzone should be characterized according to the second segment of the compressibility curve. The value of the coefficient of uniaxial compressibility for this part of the curve usually ranges from 10–20 to 0.01 MPa^{-1}, while for the initial part of the curve (up to the structural strength point) it is 2–3 times as low. The structural strength value may vary correspondingly from fractions of MPa to 3–4 MPa.

Studying of rock compressibility in the catagenesis zone is rather difficult because special apparatus are required capable of conveying to a sample a normal loading as high as 120–200 MPa. The most reliable data on the compressibility is obtained when samples are tested under the conditions of triaxial compression with loading values exceeding the geostatic ones. To conduct such tests, special apparatus must be used allowing one not only to create high pressures but also to regulate the temperature regime in testing. Therefore, such experiments known from literature should be considered as unique. And for this reason, information on the compressibility of clays in the catagenesis zone is rather scarce. The available from literature values have been obtained under the lesser than geostatic loading, and so they characterize rock compressibility on the unloading branch of the curve. The values of the coefficient of uniaxial compressibility thus obtained are too low, although there may be exclusion from this rule for highly lithified clay varieties (argillites, siltstones, cemented argillite-like clays), as the unloading causes a sharp increase in the microfracturing of these rocks that may cause their enhanced compressibility under repeated compression on the unloading branch of the curve.

The most interesting investigations on the compressibility of sedimentary rocks under the high-pressure conditions of triaxial compression have been conducted by M.P. Volarovich [160], V.M. Dobrynin [161], G.M. Avchan et al. [162]. The studies were carried out with due regard for the changes of formation pressure and temperature. The data obtained by these authors not only give an idea of the compressibility of clays in the catagenesis zone, but also allow one to estimate the associated changes in their physical properties (electric resistivity, elastic wave velocities), solve a number of practical problems in the calculation of hydrocarbon reserves, estimation of the screening properties of clay seals, etc.

3.4.2. UNIAXIAL COMPRESSIVE STRENGTH

The uniaxial compressive strength is determined by means of crushing, under gradually increasing loading, a cylindrical rock sample having a height 1.5–2.0 times as great as its diameter. Knowing the breaking load for the sample (P) and its cross sectional area (S), one can obtain the uniaxial compressive strength of the rock (σ_c)

$$\sigma_c = P/S \tag{3.13}$$

Besides the strength estimation, the total strain modulus (secant modulus) (E_t) and elastic strain modulus (Young's modulus) (E) are determined from the deformation curve of uniaxial compression. This is done by finding the relative elastic and total strain of a sample in its testing under the loading–unloading regime. Knowing the values of relative elastic (ε_e) and total (ε_t) strain and their corresponding stresses (σ), one can determine the elastic and total strain moduluses

$$E = \sigma/\varepsilon_e \qquad\qquad (3.14)$$

$$E_t = \sigma/\varepsilon_t \qquad\qquad (3.15)$$

The uniaxial compressive strength, as it is easily determined, is often used for the classification of clay rocks according to their strength and for the estimation of their degree of lithification. In this case, it is not only the strength value, but also the deformation behavior of the sample in the process of compression is of considerable interest. The latter characteristic is often underestimated, and so the results of uniaxial compression tests are given without an analysis of the deformation behavior of the sample. Meanwhile, it is known by experience that while the strength value of clay rocks depends both on the type of contacts and their number per unit cross sectional area of rock failure, the deformation behavior is mainly determined by the type of contacts. Therefore, an analysis of deformation curves of uniaxial compression allows a rather definite judgment on the predominant type of contacts in the rock.

The uniaxial compression of sediments from the early diagenesis subzone is not usually carried out because of the latent fluid nature of their deformation determined by high porosity of the sample and by the presence of extremely weak distant coagulation contacts. When even insignificant loading (10^2–10^3 Pa) is applied to such sediments, they lose strength and begin to flow.

Sediments from the late diagenesis subzone, when tested under uniaxial compression, behave like typical plastic systems. As can be seen in Figure 3.7, a, a linear segment on the deformation curve of such sediments is poorly pronounced. Already at small loading, the deformation passes to the non-linear

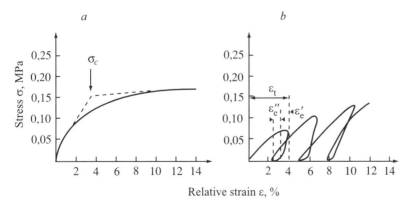

Fig. 3.7. Deformation behavior of clay sediments from the late diagenesis subzone under uniaxial compression: a – in strength testing, b – in testing under the loading–unloading regime.

stage like the viscous flow that gradually passes to the plastic flow, in which the sample is being continuously deformed under constant vertical loading while retaining its strength and continuity. Deformation of clay muds which have been relatively consolidated under natural conditions (with porosity of 40–50%) may proceed without the formation of visible fractures until relative strains of 10–15% and more are reached. Less consolidated muds continue to deform until the sample completely loses its original shape and forms a «barrel».

For this deformation behavior, the value of the sample strength (σ_c) is defined as the point of intersection of tangents to the initial and the final por-tions of the deformation curve (see Fig. 3.7, a). The determined strength value is 0.03–0.5 MPa depending on the degree of diagenetic compaction of a sample. The secant modulus may vary in this case from 0.1 to 10 MPa. With porosity being the same, the muds having more fine particle composition prove to be the strongest.

The low strength and the plastic character of deformation of sediments from the lower diagenesis subzone are caused by the predominance in them of close coagulation contacts. Relative displacement of structural elements on such contacts occurs along the boundary films of bound water, which have some structural strength and higher viscosity as compared to free water. An important feature of the coagulation contacts is their ability to restore instantly after destruction. Therefore, when clay sediment is being deformed, the num-ber of contacts in the shear zone remains unchanged, as each clay particle that loses contact in the process of displacement restores it at once with another nearest particle.

A good illustration to this is the deformation of plastic muds from the lower diagenesis subzone when they are tested under the loading-unloading regime (Fig. 3.7, b). As can be seen from the plot, the quick removal of load is accom-panied by instant reversible deformation (ε_e'), after which the height of the sam-ple grows slowly during several minutes at the expense of retarded elasticity (ε_e''). Most part of deformation, however, is classified as the irreversible resid-ual strain, whose relative part grows with the increase in the number of cycles and the value of loading in sample compression. Such deformation behavior under the loading–unloading mode is typical of clay systems with coagulation contacts between structural elements. As investigations by E.D. Shchukin and P.A. Rebinder [163] have shown, the reversible deformation (quick and retarded) in such systems is of entropy nature and is connected with the reorientation and deformation (bending) of thin platy particles, as well as with the thinning of a bound water film on coagulation contacts under loading and its restoration after load removal.

The uniaxial compressive strength of clay rocks from the early catagenesis subzone significantly increases. The porosity values range from 0.6 MPa in the upper part of the subzone to 4.0 MPa at its base. This is accounted for by the increase in number of coagulation contacts with an increase of clay density and by the appearance of stronger transitional contacts at the end of the stage.

A rectilinear portion is clearly seen on the deformation plots of such rocks (Fig. 3.8, a). At loads making up from 30 to 60% of the destructive values, the

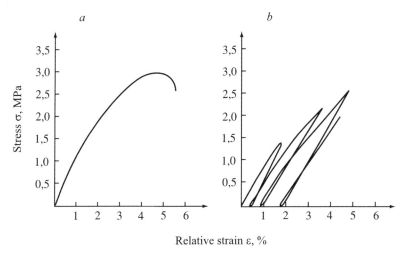

Relative strain ε, %

Fig. 3.8. Deformation behavior of clay sediments from the early catagenesis subzone under uniaxial compression: *a* – in strength testing, *b* – in testing under the loading-unloading regime.

plot $\sigma = f(\varepsilon)$ begins to flatten, thus evidencing for the beginning of viscous flow which terminates by semi-brittle destruction of the sample at a relative strain of 3–7%. The secant modulus is $10–10^2$ MPa.

An analysis of the loading–unloading plot for clays from the early catagenesis subzone (Fig. 3.8, *b*) shows that at small stresses, these clays acquire noticeable elasticity which is characteristic of solids. As the stresses grow, the role of residual strains associated with irreversible (plastic) deformation increases. The elastic modulus may be determined from the unloading curve, its maximal value being 3×10^2 MPa.

Rocks from the middle catagenesis subzone, when tested under uniaxial compression, behave like elastic solids. This is connected with their cementation and complete disappearance of coagulation contacts. Transitional and phase (cementation) contacts become predominant. Depending on the composition, the uniaxial compressive strength of the rocks varies from 3–4 to 12–15 MPa, secant modulus is $10^2–3 \times 10^2$ MPa, and modulus of elasticity is $5 \times 10^2–8 \times 10^2$ Pa. The deformation is of elastic nature almost within the entire compression range (Fig. 3.9, *a*). Only near the ultimate strength point there is a small non-linear segment that passes into brittle destruction. The relative breaking strain is 2–5%.

The change in structural bonds is reflected in the character of deformation of these rocks under the loading–unloading regime. An obvious predominance of reversible deformations as compared to irreversible residual strains is seen from the plot (Fig. 3.9, *b*).

Argillites and siltstones from the late catagenesis subzone have the maximal strength and well pronounced elastic properties. Their uniaxial compressive strength attains 50–60 MPa, modulus of elasticity is up to 10^3 MPa, secant

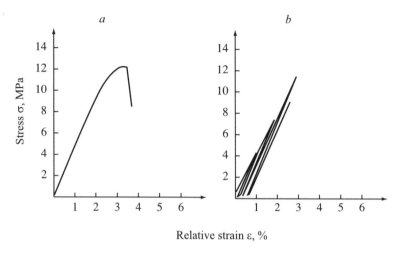

Fig. 3.9. Deformation behavior of cemented clay sediments from the middle catagenesis subzone under uniaxial compression: *a* – strength testing, *b* – in testing under the loading–unloading regime.

modulus differs insignificantly from the modulus of elasticity and attains 8×10^2 MPa. The deformation of argillites and siltstones within the whole stress range up to their destruction is practically linear (Fig. 3.10, *a*). The relative breaking strain is 1–3%. On the loading–unloading plot (Fig. 3.10, *b*), almost all the deformations are of reversible nature. The small residual strain is obviously to be attributed to the closing of microfractures.

3.4.3.   SHEARING STRENGTH

The shearing strength is characterized by the ultimate value of shear (tangential) stress resulting in the displacement of one part of a rock relative to another. Shearing tests are conducted under the conditions of normal (uniaxial) loading or triaxial compression of a sample. In the first case, they are called plane shearing (direct shear test), in the second – shearing under triaxial compression (triaxial test). The technique of shearing tests is described in more detail in a number of textbooks and monographs [3].

The main indices of rock resistance to shearing are *the friction angle, or angle of shearing resistance* (φ) and *the cohesion* (C). When φ and C are known, the shearing resistance may be found from the expression:

$$\tau = \sigma \operatorname{tg}\varphi + C \tag{3.16}$$

where σ – stress normal to the shear plane, $\operatorname{tg}\varphi$–coefficient of internal friction, C – cohesion.

A study of changes in the indices φ and C with the transformation of clay rock structural bonds in the course of lithogenesis is of great scientific and

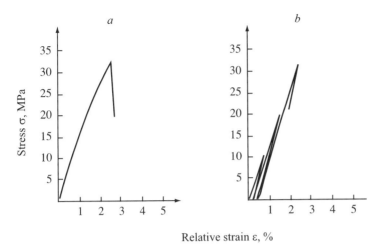

Fig. 3.10. Deformation behavior of argillites and siltstones from the late catagenesis subzone under uniaxial compression: *a* – in strength testing, *b* – in testing under the loading–unloading regime.

practical interest. Extremely valuable in this respect investigations have been done by N.N. Maslov [164], who proposed to express the shearing resistance using a trinomial rather than binomial formula:

$$\tau_{\rho w} = \sigma \mathrm{tg}\, \varphi_{\rho w} + C_{\rho w} + C_{c}$$

(3.17)

where σ – normal stress; $\varphi_{\rho w}$ – friction angle at water content w and density ρ; $C_{\rho w}$ – cohesion of a reversible character at water content w and density ρ; C_{c} – cohesion of irreversible character.

Consideration of this formula in terms of contact interactions in clays shows that, in essence, it takes into account the cohesion at the expense of both coagulation ($C_{\rho w}$) and more strong transitional and phase (C_{c}) contacts. Such a subdivision of cohesion into two components is quite justified from the theoretical and practical point of view, as the influence of coagulation and phase contacts on the strength of rocks and the character of their deformation differs significantly. In the absence of phase and transitional contacts, the cohesion of clays is determined solely by the coagulation contacts, i.e. $C \approx C_{\rho w}$. Therefore, the cohesion value depends on the density and water content of the clay, dispersion and wetting ability of its mineral constituents, degree of orientation of structural elements in the shear plane, temperature, pore pressure, etc.

In rocks with the dominance of phase contacts, cohesion is entirely determined by C_{c}, the value of which little depends on the water content, density, and composition of clay rock.

In clays with a mixed type of contacts (coagulation, transitional, and phase), cohesion is determined by both of the components, i.e. $C = C_{\rho w} + C_{c}$. In this case, at normal loading lesser than the structural strength of rock

($\sigma < P_c$), C_c plays the most important role, while at $\sigma > P_c$, the role of C_{pw} increases. So, not only the value, but also the nature of cohesion changes depending on the type of structural bonds in clay rocks.

From the same point of view one can analyze dependence of the friction angle on the character of contact interactions. The fact of principle is that friction in clay rocks with dominant coagulation contacts is not internal in its nature, but ranks as boundary friction [112, 165]. This is determined by the presence on such contacts of thin hydrate boundary films having specific properties. Shearing of structural elements relative to each other at the coagulation contacts occurs inside the boundary phase separating the structural elements and is determined entirely by its properties rather than by the properties of mineral particles. The influence of the latter may manifest itself only through their wetting ability, on which the film thickness depends.

When shearing is realized in clays with coagulation contacts, two effects are observed. The first effect is the orientation of anisometric structural elements within the shear zone in the direction of the shearing stress [166]. As a result of structural element orientation, the area of coagulation contacts noticeably increases, with a concurrent decrease in the effective stress at them, which results in the growth of the boundary liquid layer thickness and reduction of the friction at the contact. The second effect is connected with the appearance of pressure in the pore fluid during shearing, which also results in the reduction of the effective pressure at the contact. The action of both effects causes the non-linear character of the dependence of shearing resistance on the normal loading for recent sediments and weakly lithified clays.

With the appearance of transitional and particularly phase contacts, the friction of structural elements during shearing acquires the character of internal friction. Therefore, the τ value becomes dependent on the mineral composition of rocks and to a lesser extent on the pore pressure and structural rearrangement in shearing. As a consequence, the dependence of shearing resistance on the normal loading acquires a nearly linear character.

Proceeding from the above-stated, one may pass to the consideration of the peculiarities of shearing resistance in clay rocks of different lithification degree. The most complex behavior in shearing is shown by sediments from the lower catagenesis subzone. When these sediments are tested under normal loading exceeding their structural strength, the scheme of testing, i.e. the conditions of sample consolidation and drainage in shearing, becomes of importance. In testing according to the scheme of consolidated undrained shearing, dependence $\tau = f(\sigma)$ is non-linear (Fig. 3.11, a). The value of cohesion depends on the sediment density and water content and varies from 0.01 to 0.05 MPa. The value of the friction angle at small normal loading values is 12–18° and decreases to 5–10° with greater loading. This phenomenon, as has been already noted above, is explained by some peculiarities in the behavior of coagulation contacts in shearing, caused by the structural rearrangement of particles in the shear zone and by the growth of pore pressure in undrained shearing.

When sediments are tested according to the scheme of consolidated drained shearing, the shearing curve becomes more rectilinear (Fig. 3.11, b), as

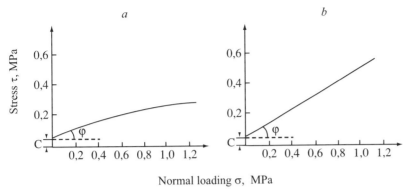

Normal loading σ, MPa

Fig. 3.11. A general view of shearing curves $\tau = f(\sigma)$ for sediments from the late diagenesis subzone: a – consolidated undrained shearing; b – consolidated drained shearing.

the influence of pore pressure is excluded in such testing. At the same time, φ grows to values sometimes as high as 20–22°.

It is noteworthy that in clay sediments having high compressibility, not only φ, but also the cohesion value change under shearing tests. The latter is connected with the increase in the number of coagulation contacts with the growth of normal pressure. This fact, together with the non-linearity of the $\tau = f(\sigma)$ dependence, provides the basis for the opinion that the Coulomb-Moor's theory is inapplicable for describing the behavior of clay sediments with coagulation contacts.

In considering the shearing strength of clay rocks from the catagenesis zone, it is important that, as the number of coagulation contacts decreases, their behavior becomes more and more similar to the shearing of solids. The rectangular character of the $\tau = f(\sigma)$ dependence becomes typical of both of the testing schemes. The explanation for this is that no noticeable change in the contact number occurs in shearing with the growth of structural bond strength and reduction of rock porosity. Besides, the effect of pore pressure and structural rearrangement of particles in the shear zone considerably decreases. Therefore, clay rocks from different catagenesis subzones differ only in the φ and C values which are mainly determined by the mineral composition of rocks and degree of their lithification. Thus, the value of cohesion increases from 0.03–0.1 MPa in the early catagenesis subzone to 0.12–0.6 MPa in the middle catagenesis subzone and to 0.12–0.8 in the late catagenesis subzone. The values of the friction angle in these subzones are 14–26°, 20–36° and 22–38°, respectively.

It is of considerable interest to examine not only the changes of φ and C depending on the dominant type of contacts, but also the peculiarities of deformation of different clays in shearing under some constant normal loading. The character of clay deformation in shearing is especially well observed in a ring shear test. Data obtained in these tests evidence that the general shape of curves $\tau = f(\sigma)$ is much similar to that of deformation curves in uniaxial compression tests. The difference is that not only the maximum (peak) strength of

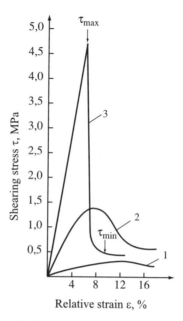

Fig. 3.12. Deformation behavior of clays in shearing at constant normal loading: 1 – clay sediments from the early diagenesis subzone, 2 – clay rocks from the early catagenesis subzone, 3 – clay rocks from the middle and late catagenesis subzones.

undisturbed structure (τ_{max}) is fixed on the shearing curves, but also the so called residual strength (τ_{min}) corresponding to the shearing resistance of rock after the destruction of its primary microstructure (Fig. 3.12).

As is seen from the plot, clay sediments from the late diagenesis subzone, as the shearing stress increases, show the non-linear deformation having an obvious plastic character (curve 1, Fig. 3.12). As the stress grows, the deformation curve gradually flattens, which evidences for the plastic flow of clay under the constant loading corresponding to the ultimate shearing strength of the structure. Some strength reduction at relative strains exceeding 12–15% is connected with the reorientation of structural elements in the shear zone and decrease of the friction angle. Consequently, a characteristic feature of clay systems with close coagulation contacts is a weakly pronounced shearing curve segment corresponding to the peak strength of sediment. This is explained by the easy recovery of coagulation contacts in deformation, so that no destruction of structures with such contacts occurs, however high be the values of relative strains.

For the tight-plastic and semi-solid clays with mixed (coagulation and transitional) contacts characteristic of the early catagenesis subzone, there is a small linear segment observed on the shearing deformation curve (curve 2, Fig. 3.12), which changes into a non-linear segment evidencing for the presence of plastic deformation. At relative strains of 8–12%, destruction of structure in the shear zone and the drop of strength occur. The residual strength value makes up no less than 30–40% of the peak one. Such a character of deformation is accounted for by the presence in such clays, besides the coagulation contacts, also of irreversibly destructible transitional contacts which cause the appearance of a linear segment at the beginning of shearing and a significant loss of strength after the destruction of structure.

The rocks of the middle and late catagenesis subzones with strong and irreversibly destructible transitional and phase (crystallization and cementation) contacts have a similar character of destruction in shearing. A characteristic feature of their deformation curve is the linear (elastic) deformation up to the brittle destruction (curve 3, Fig. 3.12) at relative strains of 4–8%. The residual strength is much lower than the peak one and makes up 10–30% of its maximal value depending on the normal loading.

3.5. RHEOLOGICAL PROPERTIES

The above-considered strength tests of clay rocks under uniaxial compression and shearing are conducted under the conditions of relatively quick growth of stress. Under natural conditions, deformation processes develop slower by several orders of values and are estimated on the geological time scale. This is connected with the slower growth of natural stresses. So, deformation of clay sediments under the conditions of their natural occurrence is mainly of rheological nature. This circumstance is of principle importance in considering the rock compaction and microfracture formation.

It is well known that even hard and very strong bodies that are destroyed in a brittle way under quick loading can be slowly deformed without breaking their continuity under the long-time action of external force. Thus, in terms of rheology, even highly lithified clays may rank according to their behavior among the plastic-viscous rather than elastic bodies. The pressure–temperature conditions, under which these rocks occur in the Earth's crust, favor such a behavior in many aspects.

The rheological processes in rocks develop under the influence of effective stresses which are concentrated at contacts and then gradually dissipate during a certain period of time passing into other kinds of energy, particularly into the mechanical one that causes displacement of structural elements along the contact. Therefore, the contacts should be considered as defects of the rock structure responsible for its rheological behavior in the same way, as dislocations determine the rheological properties of crystals. This explains the fact why rheological properties are often called the structural-mechanical properties.

The structural-mechanical properties of different structured bodies are studied by a special branch of science – the physicochemical mechanics created by the works of academician P.A. Rebinder and his disciples [167–170]. Proceeding from the theoretical principles of physicochemical mechanics, let us examine the rheological properties of clay rocks having different degree of lithification.

For recently deposited muds of the early diagenesis subzone, the most characteristic feature is a sudden loss of strength under the slightest dynamic effects (for example, weak earthquake) or due to existence of small shearing stresses. After the external force ceases to act, they manifest thixotropic restoration of

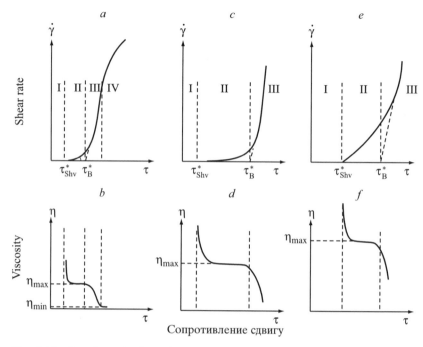

Fig. 3.13. Complete rheological curves (*a, c, e*) and dependence of viscosity on the shearing stresses (*b, d, f*): *a, b* – muds from the early diagenesis subzone; *c, d* – sediments from the late diagenesis subzone; *e, f* – clays from the early catagenesis subzone.

strength. A phenomenon associated with this is the frequent development of landslides – flows in muds at the bottom of sedimentary basins, which arise on gentle, of only few angles, slopes.

A general view of the complete rheological curve for muds is shown in Figure 3.13, *a*. As can be seen from the graph, the sediment structure, when undisturbed, can show in the field of small shearing stresses the elasticity of entropy nature (Section 1, Fig. 3.13, *a*) connected with the mutual orientation of anisometric structural elements. When stress τ_{Shv}^*, which has been called the creep limit, is reached, the mud passes to the stage of slow viscous-plastic flow of the creep type, which occurs practically without the failure of structure (segment II, Fig. 3.13, *a*) with the maximal viscosity of 10^7–10^8 Pa s. The creep develops within a certain range of values τ-τ_{Shv}^*, where the system behavior may be described according to the Shvedov's model for the viscous-plastic flow:

$$\tau - \tau_{Shv}^* = \eta_{Shv}\dot{\gamma} \qquad (3.18)$$

where η_{Shv} = differential viscosity determined by the cotangent of the curve slope angle at segment II, and γ = shear rate.

The variable (effective) viscosity at this segment is found from the equation:

$$\eta_{ef} = \frac{\tau}{\dot{\gamma}} = \eta_{Shv} \frac{1}{1 - \left(\tau_{Shv}^* / \tau \right)} \qquad (3.19)$$

After the ultimate shearing strength τ_B^* is reached, quick destruction (lique-faction) of the structure begins within a small stress range, and the system passes to the stage of viscous-plastic flow (segment III, Fig. 3.13, a) described by the Bingham's model:

$$\tau - \tau_B^* = \eta_B \dot{\gamma} \qquad (3.20)$$

where η_B – Bingham's viscosity determined by the cotangent of the segment III slope angle; τ_B^* – Bingham's ultimate shearing strength that may be considered as the ultimate strength of the structure.

Further deformation of the system (segment IV, Fig. 3.13, a) has the character of flow with constant minimal plastic viscosity of 10–30 Pa s. In this condition, the sediment is able to spread like a viscous fluid.

The complete rheological curve can also be shown as a dependence of the effective viscosity on the shearing stress (Fig. 3.13, b). The value η_{max} on this plot corresponds to the viscosity of muds with undisturbed structure, and η_{min} – to the viscosity with completely destroyed structure.

The rheological behavior of muds can be explained proceeding from the peculiarities of the distant coagulation contacts common in such sediments. Due to the small strength of bonds between structural elements, which is determined by the second potential minimum, the muds are stable only under small external stresses not exceeding the Bingham's ultimate shearing strength. In this condition, the rate of restoration of distant coagulation contacts corresponds to the rate of their destruction, and deformation has the character of creep without the loss of strength by muds. However, when τ_B^* is reached, the balance of the system becomes disturbed: there occurs an avalanche-like destruction of structural bonds which is not compensated by their restoration. As a result, with even a slight increase in the shear rate, the system passes to the condition of utterly destroyed structure and begins to flow with minimal viscosity. But this condition turns to be reversible: when the shear rate decreases or shearing ceases at all, the process of thixotropic structure formation becomes predominant over the process of its destruction. Therefore, some time after the external force ceases to act, the strength of muds becomes restored.

The rheological properties of clay sediments from the late diagenesis sub-zone are determined by the presence of close coagulation contacts in them. A characteristic feature of these deposits is their susceptibility to stable viscous-plastic deformations within a wide range of stresses exceeding the creep limit (Fig. 3.13, c and d). The plastic deformation proceeds with constant maximal plastic viscosity of 10^8–10^{12} Pa s. Neither destruction nor strengthening of the structure is observed in this case, as there occurs reversible restoration of all the destroyed coagulation contacts. Only when the ultimate shearing strength is reached, the rate of deformation sharply increases, the process of structure destruction begins to surpass the rate of its restoration, and sediments may show the signs of viscous destruction.

Clays from the early catagenesis subzone behave in rheological testing much in the same way as the sediments from the late diagenesis subzone, for the coagulation contacts are partly preserved in them. The difference is that, as transitional contacts appear in these rocks, the creep limit value τ^*_{Shv}, below which the rock shows elastic properties, increases (Fig. 3.13, e, d). When τ^*_{Shv} is exceeded, the viscous-plastic deformations develop in the rock with the constant maximal plastic viscosity of 10^{12}–10^{14} Pa s, i.e. almost by two orders of values greater than in plastic sediments of the late diagenesis subzone. With further growth of the shearing stress ($\tau > \tau^*_B$), the rate of structure destruction quickly increases, while viscosity decreases, which results in the semi-brittle destruction of rock.

Clay rocks from the middle and late catagenesis subzones, having dominant transitional and phase contacts, are elastically deformed within almost the entire loading range under usual conditions. However, under the conditions of long-time stress action measured on a geological time scale and at elevated temperature, they can, as already mentioned, show plastic deformations, being deformed with very high viscosity of 10^{15}–10^{18} Pa s. With the growth of rock dispersion and temperature, the probability of such a deformation mechanism increases.

The influence of temperature on the deformation behavior is manifested first of all in that, as the temperature grows, the local stress concentrations decrease at the expense of viscous-plastic displacement of structural elements, which lessens the probability of the formation of microfractures and their growth in the course of deformation.

The plastic deformation of cemented rocks in the catagenesis zone should be considered as not just mechanical, but also physicochemical process. A great influence on its development is exerted by the phenomenon of adsorptional reduction of the strength of solids, known in literature as Rebinder's effect [149]. This effect manifests itself most actively when a liquid phase interacts with a solid body in the presence of mechanical stresses. The basis of the effect is the phenomenon of a decrease in the specific free surface energy of solids as a result of liquid phase adsorption leading to the easier plastic deformation of this body (adsorptional plasticizing).

3.6.   GENERAL REGULARITIES IN THE FORMATION OF CLAY PROPERTIES IN THE PROCESS OF LITHOGENESIS

3.6.1.   EARLY DIAGENESIS SUBZONE

Formation and transformation of the composition and primary microstructure of sediments begin in the process of sedimentogenesis and continue at each stage of lithogenesis. In the coastal and shallow-water zones of basins, sediments rich in pelitic and sand particles accumulate. The formation of their primary structure proceeds by means of heterocoagulation and results in the origination of large-honeycomb microstructure whose framework is made up of microaggregates and individual clastic grains. In the deep-water and hydrodynamically quiet zones of sedimentary basins, more homogeneous sediments with small-honeycomb structure and porosity amounting to 85–90% are formed. The dominance of honeycomb structure in young sediments has thermodynamic nature and is associated with the tendency of a fine grained system to lower its surface energy. The honeycomb structure allows doing this with the minimal concentration of a solid phase in the volume being structured.

The connection between individual particles inside the microaggregates in a newly formed sediment is principally realized through the close coagulation contacts, whereas the connection between aggregates is dominated by the distant coagulation contacts, which form between relatively large structural elements under the conditions of straitened coagulation. The attraction force of particles at a single contact of two micron-size particles is no more than 5×10^{-10} N for the distant coagulation contact and 10^{-8} N for the close contact [38, 51, 169, 171].

The rather low energy of molecular forces operating at coagulation contacts and the small number of contacts in the honeycomb structure cause the extremely low strength of young sediments and their specific behavior under deformation. At insignificant shearing stresses they are able to manifest viscous-plastic deformation of the creep type without structure destruction with the ultimate viscosity of 10^7–10^8 Pa s. With further growth of stress or under insignificant dynamic effects, the avalanche-like destruction of coagulation contacts and loss of structural coherence occur: the muds begin to flow like viscous fluids. Therefore, young sediments are considered to have the latent-fluid consistence. After the external force ceases to act, they are able to show quick (thixotropic) restoration of their strength.

In newly deposited sediments, 80–95% of their volume is occupied with water. Water content of the sediment may be as high as 150–350%. The bulk of water (75–80% of the total quantity of water contained in the sediment) represents free water that fills large (10–100 μm) and middle-sized (1–10 μm) intermicroaggregate micropores-cells and partly also fine (0.1–1.0 μm) intramicroaggregate micropores. Physically bound water makes up 20–25% of the total water mass in the sediment and forms boundary layers of bound water of several hundredth, sometimes tenth fractions of μm in thickness on the surface of clay minerals. It obstructs completely the most numerous interparticle and intramicroaggregate ultramicropores (<0.1 μm) and significantly reduces the «live» cross section of fine micropores. In spite of this, permeability of young sediments is rather high and equals 0.3–2 mD (Table 3.1).

Practically immediately after the accumulation, the process of sediment compaction begins. At the very beginning of diagenesis, the compaction proceeds at the expense of physicochemical processes tending to reduce the free surface energy of sediments. This is manifested in the spontaneous condensing of sediment structure as a result of syneresis and colloid aging. When a sediment of some tens of centimeters or few meters in thickness is accumulated, along with the physicochemical processes, also the geostatic compression becomes apparent. Under the weight of overlying deposits, slow relative displacement of structural elements and their closer packing occur in the sediment. The displacement occurs mainly along the weak distant coagulation contacts between microaggregates and is accompanied by a decrease in sediment porosity due to compression of large microaggregate pores. The deformation during compaction is of viscose-plastic (creep) nature and is controlled by the rate of free pore water squeezing-out. Therefore, during the compaction of such sediments, continuous redistribution of the total stress between the pore stress and the effective stress proceeds.

As a result of compaction, the sediment porosity decreases to 60–75% already by the end of the early diagenesis substage. Water content drops to 45–95%, although its value remains higher than the liquidity limit (Table 3.1). The content of large intermicroaggregate micropores decreases.

3.6.2.   LATE DIAGENESIS SUBZONE

In the late diagenesis subzone, the geostatic compaction continues to be the main factor of further sediment transformation. The compaction is accompanied by a gradual growth of structural bond strength due to the strengthening of individual contacts and increase in their number. The distant coagulation contacts gradually convert into the contacts of close coagulation. The sediments acquire fluid-plastic and soft-plastic consistence and begin to manifest well pronounced plastic properties. Their uniaxial compressive strength attains 0.03–0.5 MPa. Deformations are mainly of a residual (plastic) character, secant modulus is 0.1–10 MPa. In slow deformation, the sediments show viscose-plastic flow (creep) with undisturbed structure viscosity of 10^8–10^{12} Pa s (Table 3.1).

The shearing strength of sediments is determined by specific properties of the coagulation contacts, where friction is of the border type in its nature. This causes deviation of sediment behavior in shearing from the classical Coulomb-Moor's theory. Its important feature is the curvilinear shape of the shearing curve evidencing for the variation of φ and C values with the growth of normal loading. The principal cause of such a behavior is the change in the number of contacts and the structural rearrangement of particles in the process of shearing, as well as the pore pressure growth in shearing. The specific nature of shearing determines rather small values of its parameters: the cohesion value (C) varies within a range of 0.01–0.05 MPa, and φ ranges from 5 to 18°.

During the entire late diagenesis substage, sediments continue to be intensely compacted. The compaction mechanism remains the same as in the early diagenesis subzone. Compressibility continues to be high enough. This is evidenced by the coefficient of compressibility values ranging from 20 to 0.01 MPa^{-1}. Sediment porosity at the end of the late diagenesis substage becomes close to 35–45%. The largest contribution to the pore space volume is made by small and fine intermicroaggregate micropores. The reduction of total porosity and increase in the content of small and fine micropores in the pore space structure results in the reduction of permeability to 10^{-1}–10^{-3} mD.

Water content of sediments at the end of the late diagenesis substage becomes close to the liquid limit or somewhat lower than it and is 30–45%. The remaining water ranks as physically bound according to its state. Thus, by the end of diagenesis, the first stage of clay sediment dehydration terminates resulting in the removal of practically all free water from the sediment. The total mass of squeezed-out free water makes up 70–80% of the moisture originally stored in the sediment.

3.6.3.   EARLY CATAGENESIS SUBZONE

After the passage of clays into the early catagenesis subzone, the coagulation contacts continue to be dominant in them. However, already in the upper part of the subzone and especially in its second half, the process of their gradual transformation into the transitional-type contacts begins, with the strength of the latter being by an order of values greater than that of the coagulation contacts. The transitional contacts form in two ways. The first way is the thinning (squeezing-out) of hydrate films at the coagulation contacts, with the ionic-electrostatic bonds being formed between particles, similar to those existing between structural layers in layered silicates. The second way is associated with the beginning of clay cementation and differentiation on a very restricted area of a new phase – the embryo of future cementation contact.

A change in the character of structural bonds leads to a gradual increase in the strength of clays and to the alteration of their behavior (Table 3.1). Rock consistence within the early catagenesis subzone changes from tight-plastic to semi-solid. In the uniaxial compression and shear testing, a linear segment appears at the initial portion of deformation curves, providing evidence for the increasing role of elastic strain. With the growth of loading, the deformation acquires the character of viscose-plastic flow terminated by semi-brittle destruction. The uniaxial compressive strength of clays attains 0.6–4.0 MPa. Secant modulus varies within a range of $10–10^2$ MPa, modulus of elasticity is about 3×10^2 MPa. With the rise of transitional contacts and true (internal) friction at the particle contacts, the shear parameters increase. The cohesion increases to 0.03–0.1 MPa, and the friction angle reaches 14–26°. The values of undisturbed structure viscosity increase by almost two orders and are $10^{12}–10^{14}$ Pa s.

The compaction that develops intensely at the beginning of the early catagenesis subzone becomes noticeably slower towards the base of the subzone in connection with the strengthening of structural bonds and beginning of cementation. One may suggest that near the boundary of the early and middle catagenesis substages, the stage of free geostatic rock compaction, that continued until strong structural bonds were formed in the rock, comes to an end and the stage of impeded compaction begins (Table 3.1). Rock porosity towards the base of the early catagenesis subzone attains the values of 16–25%. Fine and small intermicroaggregate micropores are dominant (in volume) in the pore space structure.

The continuing compaction of sediments in the early catagenesis subzone causes their further dehydration and a decrease in water content which occurs mainly at the expense of squeezing-out of loosely bound water, and in the first place osmotic, water. The mechanism of this process is the squeezing-out of hydrated ions of diffuse layers from ultramicropores into larger pores, with their further squeezing-out beyond the sequence under the influence of hydrostatic pressure. This is opposed by the osmotic mass transfer directed towards the rocks undergoing dehydration. In general, the process of pore water squeezing-out becomes more complicated and is connected not only with purely mechanical but also with physicochemical phenomena. This causes further reduction of permeability and improvement of the screening properties of clays. Towards the base of the early catagenesis subzone, the bulk of osmotically bound water proves to be squeezed out. The weight water content becomes equal to, or slightly higher than, that of the plastic limit and is 10–18%.

3.6.4.   MIDDLE CATAGENESIS SUBZONE

In the middle catagenesis subzone, cementation of rocks and their further strengthening at the expense of gradual transitional-to-phase contact transformation become widely developed. The strength of such contacts calculated for the interaction of two micron-size particles is no less than 4×10^{-7} N. The phase contacts form at the expense of increase in the area and strength of transitional contacts having the ionic-electrostatic nature and their transformation into the areas of crystallization intergrowth of minerals, as well as through the accretion of transitional (point) cementation contacts with the formation of a new mineral phase cementing clay particles and clastic grains.

With the development of cementation, the rock acquires the properties of a solid that deforms predominantly elastically up to the brittle destruction. The modulus of elasticity and the secant modulus become close in their values and are 5×10^2–8×10^2 and 10^2–3×10^2 MPa, respectively. The strength characteristics become noticeably higher. The uniaxial compressive strength attains 3–15 MP·a, the cohesion in shearing is 0.12–0.6 MPa, the friction angle is 20–36° (Table 3.1).

In spite of the strength increase, the deformation of clays under the conditions of high temperature and pressure continues to be viscose-plastic and develops as a creep of undisturbed structure with extremely high viscosity of 10^{15}–10^{18} Pa·s. On this basis, one can suggest that at the beginning of the middle catagenesis subzone, rocks continue to be compacted without breaking their continuity and formation of microfractures.

In the middle of the middle catagenesis substage, the second stage of water dehydration – removal of loosely bound water – comes to an end. Water content of rocks becomes close to the maximal hygroscopic value and, as a rule, does not exceed 5–8%.

Further dehydration of rocks is associated with the squeezing-out of adsorptionally bound water. The mechanism of removal for this category of water is of thermodynamic nature and involves increase in the mobility of adsorbed water molecules with temperature growth and their passage into free state.

This process begins at the temperature of $65-70°C$, which most often corresponds to the middle part of the middle catagenesis subzone, and continues in the late catagenesis subzone.

The transformation of adsorptionally bound water into free water proves to be an important event in the course of lithogenesis. Associated with it is the passage from the progressive run of lithogenesis to its regressive stage. Before this process begins, the density of rocks progressively increases and their porosity and permeability decrease. Concurrently, the screening properties of clays as seals grow. In the middle part of the middle catagenesis subzone, rock permeability attains its minimal values ($10^{-3}-10^{-5}$ mD). This is caused not only by the greatly reduced pore volume, but also by the change in the pore space structure: thin intermicroaggregate micropores and interparticle and intramicroaggregate ultramicropores become predominant. The most part of the «live» cross section in these pores proves to be obstructed by adsorptionally bound water and does not participate in the filtration transfer of moisture. Another important circumstance is that, up to the middle of the middle catagenesis substage, the deformation of rocks under compaction develops according to the viscose-plastic mechanism and does not result in the formation of microfractures. Under these conditions, the filtration permeability is present only at high hydrodynamic pressures. A favorable situation is created for the formation of anomalously high pore pressure (overpressure).

As the transformation of adsorptionally bound water into free water starts, the regressive stage of lithogenesis begins, which brings to deterioration of the screening properties of clay sequences. With its development, the processes of dissolution and authigenic mineral formation, including hydromicatization of montmorillonite and mixed-layer varieties, become activated; the structural rearrangement of rock proceeds with an increase in its structural heterogeneity; microfractures begin to develop. If dissolution processes are intense, anomalous (decompacted) horizons may form within the rock sequence.

A number of phenomena arising in the process of tightly bound-to-free water transformation are of principle importance. These are: (a) protonization of forming free water, which causes its high dissolving power; (b) decrease in the effect of adsorptional reduction of the strength of solids (Rebinder's effect), leading to the increase in rock brittleness; (c) enlargement of the «live» pore cross section with disappearance of boundary layers of adsorbed water molecules, which causes an increase in the pore permeability of rocks.

So, within the middle catagenesis subzone, the alterations occur in rocks, which are of great importance in estimating the clay rocks as seals. In the upper part of the subzone, the lithogenesis conditions favor the formation of the best isolating properties of clay rocks, as well as the formation of overpressures. When an isotherm of $65-70\,°C$ is crossed, the picture begins to change. Heterogeneities arise in the rock structure in connection with the progress of dissolution and mineral neocrystallization, which results in the formation of stress concentrators. The increase in rock brittleness causes the origination of microfractures. The latter, together with the «opening» of ultramicropores and fine micropores due to disappearance of boundary layers of bound water, results in the increase of clay rock permeability to 10^{-2}–10^{-3} mD and deterioration of their screening properties.

3.6.5.   LOWER CATAGENESIS SUBZONE

In this subzone, the processes resulting in the clay seal degradation continue to develop. With the beginning of argillization, the process of «intergrowth» of clay particles inside microaggregates goes on to form large polycrystalline overgrowths of platy shape, which further enhances the heterogeneity of rock structure. The most weakened zones prove to be the contact boundaries of polycrystalline overgrowths, along which microfractures arise. As the degree of argillization increases, the number of microfractures also increases, leading to further reduction of screening properties.

Despite the degradation processes, the strength of rocks remains high and in a number of cases increases at the expense of the phase contact area growth. The uniaxial compressive strength of argillites and siltstones may be 5–60 MPa. The modulus of elasticity and the secant modulus acquire practically similar values equaling to 10^3 and 8×10^2 MPa, respectively. The cohesion in shearing attains 0.12–0.80 MPa, and the friction angle is 22–38° (Table 3.1).

The compaction mechanism is changed again. The geostatic compaction becomes highly impeded. Crystallization and partial recrystallization of rock substance, in the process of which new mineral phases filling the pore space are formed, begin to play a more and more important role. Therefore, rock compaction in the late catagenesis subzone proceeds due to the combined action of geostatic pressure and crystallization processes (the third stage of the lithogenetic compaction process). Towards the subzone base, the rock porosity decreases to 2–4%, but at the same time the fracture void space increases at the expense of microfractures.

3.6.6.   METAGENESIS ZONE

With the progress of metamorphization in the metagenesis zone, the processes of rock substance recrystallization intensify. Active rearrangement of structure proceeds with development of schistosity and cleavage.

The main type of void space becomes that of microfractures whose development continues up to certain critical values, after which their growth stops and «healing» by newly formed mineral substance begins.

Small changes in the density and porosity of rocks, which occur in this zone, are connected exclusively with the processes of rock recrystallization.

The above-considered regularities in the formation of clay seal properties in the course of lithogenesis are based on the generalization of available experimental data and theoretical premises ensuing from the physicochemical mechanics of rocks. The conclusions drawn are of a generalized character and should be adapted to each individual oil and gas basin proceeding from the peculiarities of clay sediment composition, geological history of the basin, and pressure–temperature conditions of the lithogenesis process.

Part II

CHARACTERISTIC OF THE FACIES TYPES OF CLAY SEALS

4

Depositional Environments of Clay Seal Formation

4.1. THE CYCLIC STRUCTURE OF SEDIMENTARY BASIN SEQUENCES

A clay seal in a sedimentary oil- and gas-bearing basin represents a geologic body corresponding to a certain stage in the cycle of sedimentation.

It has been shown by many researchers [148, 172] that the formation and quality of clay seals in oil- and gas-bearing basins are closely associated with the character of accumulation of oil- and gas-prone organic matter and controlled by their depositional environments and general geological character of basin development.

The primary depositional environments in which a sediment forms determine its initial mineral composition, granulometry, and geochemical features.

Sedimentary sections of basins located within both ancient platforms and stabilized uplifted portions of young platforms, as well as within alpine geosynclinal belts and activated subsiding zones of epi-Hercynian platforms and orogens, have a well pronounced cyclic structure [107, 108, 173–177]. The latter is manifested in the alternation of sediments of alluvial and near-shore marine genesis accumulated during a regressive phase of basin development and those of true marine genesis accumulated during a transgressive phase.

Within the well-known West-Siberian basin [173], periods of transgression and regression, characteristic of the basin as a whole, manifested themselves throughout its area as an alternation in the section of sand-silt and clay deposits of continental and marine genesis. A characteristic feature of the basin is the significant dominance in it of marine deposits, which is highly favorable for petroleum accumulation both in terms of facies that control the type of organic matter and geochemistry of environment, and from the standpoint of rock composition that determines the quality of reservoirs for hydrocarbon accumulations and of their seals.

A number of researchers [107, 108] point out that formation of the thickest clay sequences, playing later the role of screens, corresponds to the maximum of transgression, when marine basin occupied the largest area and the composition of clay material brought from land was dominated by fine grained minerals with a swelling crystalline structure.

An analysis of sedimentary basin sections shows that at their base, they are composed of river-bed alluvial sediments, which change upwards into sediments of delta, shallow-water marine zone and shelf and are crowned with clay sediments of deep-water marine environments [107, 173, 178, 179]. The sedimentary cycles show periodical recurrence within a sedimentary basin section, with a gradual increase in the share of transgressive deposits up the section. A detailed study of sedimentary series representing complete sedimentary cycles reveals the presence of fine grained, low-permeability and impermeable clay units serving as seals which separate the sand-silt sequences – potential reservoirs for oil and gas.

Different researchers [178, 180–182], when studying the periodicity of sediment deposition in the Earth's history, have revealed transgressive epochs expressed within terrigenous sequences through the formation of homogeneous clay units. Among the latter, there are regional seals with the best properties, which have formed in deep-water marine environments. Within carbonate sequences, the most prospective zones for seal formation are regressive series – the intervals where deep-water carbonate sediments are followed by near-shore marine clays.

For the West-Siberian oil- and gas-bearing basin [107], a comparison has been made between the maximums and minimums of sea extension during Mesozoic time and the planetary waves of transgressions and regressions [178]. A well pronounced synchronism has been established (Fig. 4.1). It is noteworthy that the Upper-Jurassic and Upper-Cretaceous sediments of West Siberia form regional seals screening numerous oil and gas fields. A similar rhythm has been found to exist on the Turanian and Scythian platforms, where sediments of transgressive complexes synchronous with planetary transgressions form regional seals.

Thus, a geological basis for the revealing of regionally persistent clay seals is the follows: recognition of the rhythmic structure of a sedimentary basin

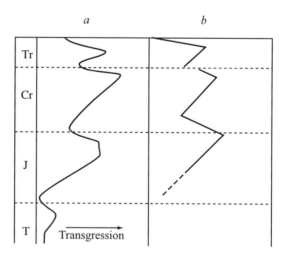

Fig. 4.1. Rhythmic occurrence of the planetary wares of transgression and regressions in the Mesozoic–Cenozoic (*a*), as compared with the corresponding waves of transgressions and regressions in the West-Siberian lowland (*b*) [107].

section, definition of the stratigraphic position of transgressive and regressive rock complexes, knowledge of paleoclimatic zonation and time of humid or arid epoch development in geological history. Identification of sedimentary series representing a complete sedimentary cycle makes it possible to outline the stratigraphic intervals within the section, where low-permeability rocks serving as seals are developed, whereas the separating them sand-silt sediments represent petroliferous complexes [181].

4.2.   SOURCE PROVINCES

Sedimentogenesis begins with the mobilization of sediment matter realized through hypergene processes, which is followed by its transportation and accumulation in the form of deposit within a sedimentation basin. Continental blocks adjacent to the sedimentation basin are the source areas providing it with detritus and phytogenous organic matter. A quiet tectonic regime and humid climate in source areas favor the wide development of complete profiles of weathering crusts on sedimentary, metamorphic, and igneous rocks [183]. In the hypergenesis zone, within crusts of weathering, conditions arise for a progressive transformation of the indicated rock complexes through the processes of oxidation, hydration, dehydration, replacement, and hydrolysis of rock-forming minerals.

The main factor determining the character of weathering is the climate, and inside the climatic zones – the relief of territories. Relief determines the conditions of water exchange, which are very important in rock disintegration. These factors, as well as the composition and fabric of initial rocks, influence the composition of weathering products [77, 184, 185].

I.I. Ginzburg [186] has shown that rocks of different composition give different products of disintegration. He gave a general scheme of changes in the weathering profiles of acid, basic, and ultrabasic rocks depending on the climate. It is seen from the scheme (Fig. 4.2) that weathering crusts of the same rocks occurring within the same region (the South Urals), but formed in different climatic zones have different compositions.

The weathering products, depending on the type of processes controlling their formation, divide into two groups: (1) formed under the influence of physical and physicochemical processes and (2) formed under the influence of chemical processes (hydrolytic decomposition).

According to the time of formation, the zones of a weathering crust profile may be both coeval and of different ages. In ancient crusts of weathering, only the lower horizons of the profile are preserved, while the upper horizons are in most cases secondary, superposed. Figure 4.3 shows the complete sections of ancient weathering crusts from the Turgay trough and Ukrainian shield. In both cases, the zoned structure is clearly seen. Developed on a bedrock is a horizon of montmorillonite clays followed upwards by hydromica clays, which change into chlorite and kaolinite clays [187].

Formation of clay minerals in a weathering zone occurs in successive stages. Originally, clay minerals of complicated chemical composition arise, whose

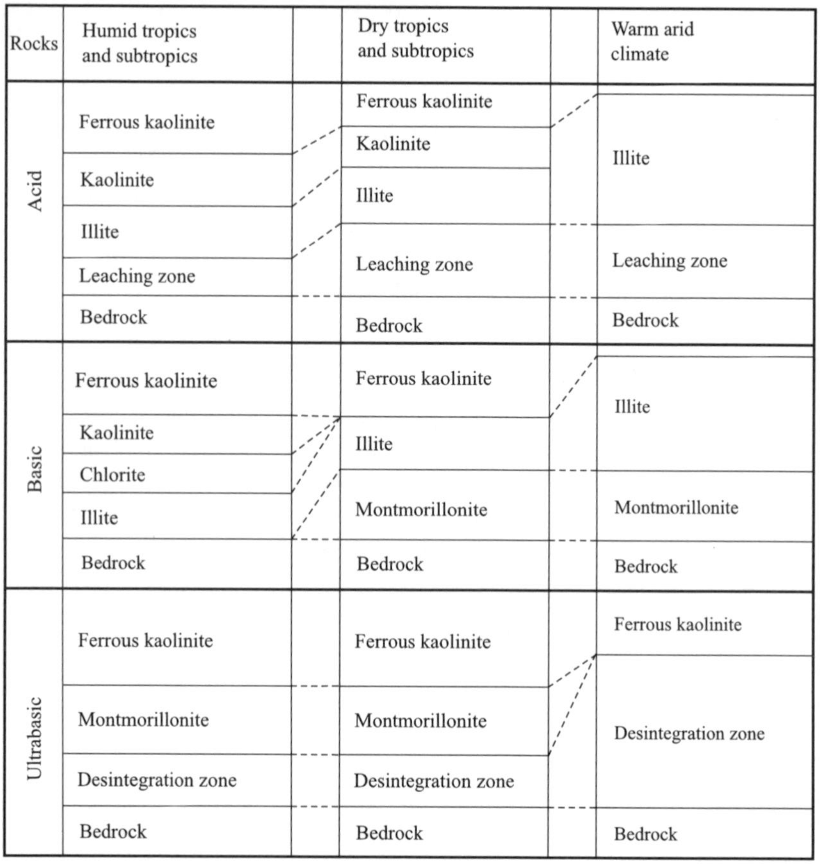

Rocks	Humid tropics and subtropics	Dry tropics and subtropics		Warm arid climate
Acid	Ferrous kaolinite	Ferrous kaolinite		Illite
		Kaolinite		
	Kaolinite	Illite		
	Illite			
	Leaching zone	Leaching zone		Leaching zone
	Bedrock	Bedrock		Bedrock
Basic	Ferrous kaolinite	Ferrous kaolinite		Illite
	Kaolinite	Illite		
	Chlorite			
	Illite	Montmorillonite		Montmorillonite
	Bedrock	Bedrock		Bedrock
Ultrabasic	Ferrous kaolinite	Ferrous kaolinite		Ferrous kaolinite
	Montmorillonite	Montmorillonite		Desintegration zone
	Desintegration zone	Desintegration zone		
	Bedrock	Bedrock		Bedrock

Fig. 4.2. Changes in the weathering profiles of acid, basic, and ultrabasic rocks depending on the climate.

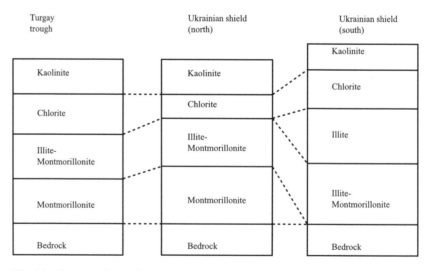

Fig. 4.3. Cross-sections of ancient crusts of weathering.

disintegration results in the formation of minerals having simpler compositions. The end products of disintegration are clay minerals which are stable in the upper horizons of weathering crusts: kaolinite, chlorite, illite, and vermiculite. A characteristic feature of the lower weathering crust horizons is the presence of large quantities of mixed-layer, predominantly illite-montmorillonite minerals and montmorillonite. Among them, the swelling layered silicates, such as illite-montmorillonite and montmorillonite with a high content of swelling component, develop preferably after magmatic complexes, and the non-swelling minerals – illite, ferrous-magnesium chlorite, and kaolinite – form mainly through the transformation of ancient sedimentary and metamorphic rocks.

Accumulation in sedimentary deposits of the montmorillonite-group and mixed-layer minerals with a great amount of swelling layers occurs only during quiet tectonic epochs favoring the long-time stay of rocks in the hypergenesis zone and the formation of a complete weathering crust profile [188]. During the epochs of intense tectogenesis, sedimentary basins are supplied with poorly reworked material considerably enriched in illite and chlorite.

So, a necessary condition for the mobilization of clay matter in source provinces is a stable tectonic regime and humid climate, which assists in the development of hypergene processes with the formation of complete weathering crust profiles containing the horizons of mixed-layer varieties and montmorillonite at their bases.

4.3. EVOLUTION OF SEDIMENTATION BASINS

A depositional environment may create favorable conditions for the formation of various seals differing in their lithology and properties. The thickness of seals depends on the amount of downwarping of the sedimentation basin bottom under the conditions of compensated sediment accumulation and increases towards the center of the basin. The maximal thicknesses tend to be located in the areas of relatively deep-water formations [107, 189].

It is known that the majority of the largest oil and gas fields have been formed in sediments deposited within a shelf and adjacent part of sea [1, 190]. It is also known that clay components, together with carbonaceous organic matter, were supplied from land to the sedimentation basin and deposited within three main bathymetric regions: (a) shallow-water coastal and shelf areas, (b) relatively deep-water part of shelf, and (c) deep-water shelf and adjacent sea. This led to the accumulation of oil- and gas-producing organic matter as well as of clay masses that were to become seals in future.

Differences in the conditions of sediment accumulation within various parts of shelf and marine basin proper determine the formation of clay seals with different screening properties, which are closely associated with the primary genetic characteristics of sediments. It is the depositional environment that determines such an important feature of clay seals as their heterogeneity. Thus, seals having the most homogeneous lithological composition with minimal contents of sand-silt interlayers form in those portions of marine sedimentation basins, which

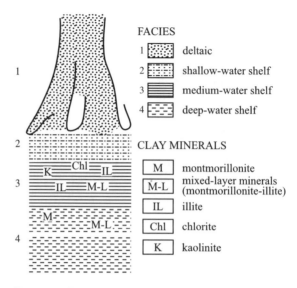

Fig. 4.4. A sedimentary oil- and gas-bearing basin on a plan view. Facies of: 1 – deltaic, 2 – shallow-water shelf, 3 – medium-depth shelf, 4 – deep-water shelf. Clay minerals: M – montmorillonite, ML – mixed-layer (montmorillonite-illite), IL – illite, CHL – chlorite, and K – kaolinite.

are located far from the shoreline (deep-water shelf and adjacent part of sea) and characterized by a quiet hydrodynamic regime of bottom waters (Fig. 4.4).

The number and thickness of permeable interlayers in seals increase towards the shallow-water coastal areas and paleodeltas of rivers. As the sedimentary basin depths become shallower, the clay sequences become less persistent in thickness, the contents of sand and silt interlayers increases, the clays become less fine grained, and their structure becomes less homogeneous [191].

Based on the lithological-facies analysis and analysis of conditions governing the formation of clay sequences in specific Mesozoic oil- and gas-bearing sedimentary basins of the West-Siberian, Scythian, Turanian, and Russian platforms, West Africa, eastern and western Atlantic framework, a lithological-facies model of seal formation can be proposed. The model describes the formation of seals with different screening properties depending on the paleogeographic conditions of sedimentation and general geological evolution of the sedimentary basin and adjacent continental area.

The proposed model includes three stages of sedimentation basin evolution: the beginning of transgression, the middle of transgression, and the maximum of transgression (Fig. 4.5).

4.3.1.   BEGINNING OF TRANSGRESSION

At the initial stage of oil- and gas-bearing basin development, a sedimentation basin is supplied from the continent with a great amount of clay material formed in the hypergenesis zone and derived from the upper horizons of weathering crusts, together with abundant organic matter of the humus type. Sedimentary

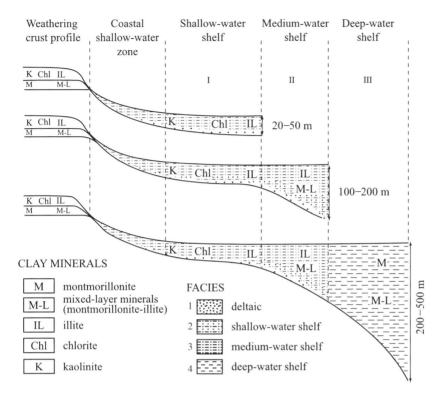

Fig. 4.5. A scheme of sedimentation basin evolution. Stages of evolution: I – beginning of transgression, II – middle of transgression, and III – maximum of transgression.

oil- and gas-bearing basins, that accumulate this matter, develop under certain physiogeographic, climatic, and tectonic conditions.

A necessary condition for the formation of a basin is its active downwarping, as well as development of certain physiogeographic processes. These processes affect the formation of a river system and subaerial delta plain. The river system, including a drainage basin, alluvial valley, and delta plain, serves as a supplier of clay material from the source area to the accumulation basin. The subaerial delta plain makes part of the shallow-water coastal zone, where currents and wave processes are active (Figs. 4.4 and 4.5).

Sediment material is supplied from the zone of weathering to the sedimentation basin. The material which settles in the zone of shallow-water coastal and shelf sedimentation, under the conditions of shallow depths (25–50 m) and active water dynamics caused by the activity of deltas, near-shore currents, and tides, is represented by sand-silt particles together with humus-type organic matter (Fig. 4.5, basin evolution stage I). Concurrently there settle, due to the conditions of mechanical sedimentary differentiation, the largest clay particles and their aggregates predominantly of kaolinite, chlorite, and illite compositions, derived from the upper horizons of weathering crusts (Fig. 4.4). In the exchange complex of clay particles, bivalent cations (calcium and magnesium) prevail, which promotes the formation of large clay aggregates. The forming sediments

Table 4.1. Classification of clay seals according to their depositional environments and screening

Depositional environments	Mineral composition	Prevailing particle size, μm	Lithologic homogeneity	Average content of sand-silt material	Content of carbonate material	Type of organic matter
Deep-water shelf and adjacent part of sea. Depths 300–500 m. Quiet hydrodynamic regime	Montmorillonite	<0.1	Homogeneous	Up to 10%	>1%	Sapropel
Deep-water shelf. Depths 200–300 m. Quiet hydrodynamic regime	Montmorillonite, mixed-layer varieties (illite-montmorillonite)	0.1–0.2	Homogeneous	10–20%	>1%	Sapropel
Middle shelf. Depths up to 200 m. Relatively quiet hydrodynamic regime	Mixed-layer (illite-montmorillonite), illite	<0.3	Largely homogeneous	20–30%	Up to 1–2%	Sapropel
Peripheral delta area. Depths up to 100 m. Weak influence of turbidity currents	Illite, mixed-layer (illite-montmorillonite)	0.3–1	Heterogeneous, interbedding with siltstones	30–40%	Up to 2–3%	Sapropel–humus
Shallow shelf. Water depths up to 50–70 m. Relatively active dynamics	Illite, kaolinite, chlorite	0.5–3	Heterogeneous, interbedding with siltstones, sandstones	40–50%	Up to 5%	Humus, admixture of sapropel
Near-shore agitated shallow water. Active dynamics. Depths up to 30–50 m	Kaolinite, illite, chlorite	0.5–5.0	Heterogeneous, interbedding with siltstones, sandstones	50–70%	Up to 5%	Humus
Near-shore highly agitated shallow water, very active dynamics. Depth 20–30 m	Kaolinite, chlorite, illite	0.5–10	Heterogeneous, interbedding with siltstones, sandstones	70–80%	Up to 5%	Humus

contain sand and silt material in quantities from 40–50% to 70–80% (Table 4.1). Active water dynamics causes the cross bedding of sediments accentuated by the oriented arrangement of organic matter and sand- and silt-size grains.

Thus, the shallow-water coastal and shelf environments are characterized by the formation of rather thin (3–5 m) clay layers with admixture of sand-silt material, which are composed mainly of kaolinite with subordinate chlorite and illite.

4.3.2.   MIDDLE OF TRANSGRESSION
Further development of transgression occurs with continuous downwarping of the sedimentation basin. Clay material is deposited in the medium-depth (100–200 m) environments on shelf and in the peripheral part of delta (Fig. 4.5, basin evolution stage II). Rather homogeneous clay and low-silt (containing

properties. (Table 4.1. continued)

Ratio of exchange cations Na^+/Ca^{2+}	Capability of fracturing	Permeability to gas, absolute, mD	Diffusion coefficient, cm^2/sec	Permeability in situ to		Seal thickness	Class
				Gas	Oil		
6–12	Very low	10^{-5}	10^{-10}–10^{-9}	Impermeable	Impermeable	Few – tens of meters. Reliable screen for oil, gas	I
4–12	Very low	10^{-5}	10^{-10}–10^{-9}	Impermeable	Impermeable	Few – tens of meters. Reliable screen for oil, gas	II
3–5	Low, some microfractures present	10^{-4}	10^{-8}–10^{-7}	Diffusionally low-permeable	Impermeable	Tens of meters, unreliable screen for gas, reliable – for oil	III
2–4	Average, microfractures present	10^{-3}	10^{-7}–10^{-6}	Diffusionally permeable	Impermeable	Tens of meters, reliable screen for oil	IV
1–3	Average, microfractures present	10^{-2}	10^{-6}–10^{-5}	Diffusionally permeable, filtrationally permeable	Diffusionally low-permeable	Tens of meters, reliable screen for oil	V
~1	Enhanced, microfractures present	10^{-2}	$>10^{-5}$	Filtrationally permeable	Filtrationally permeable	Does not screen oil and gas independently	VI
~1	Enhanced, microfractures present	$<10^{-1}$	$>10^{-5}$	Filtrationally permeable	Filtrationally permeable	Does not screen oil and gas independently	VII

20–40% of sand-silt grains) sediments accumulate within the peripheral part of delta surrounding the submarine delta foot; their formation is connected with the supply of fine grained material from the delta side (Figs. 4.4 and 4.5). Sediment deposition proceeds under the conditions of weak water dynamics, which results in the formation of a relatively homogeneous and lithologically persistent sediment having horizontal bedding accentuated by layer-by-layer accumulations of fine sapropel–humus organic matter. At this stage of sedimentation basin evolution, clay components are supplied to it from the middle horizons of weathering crusts, which have undergone deeper mechanical sedimentary differentiation. This leads to the accumulation of better sorted fine grained material of predominantly illite and mixed-layer (illite-montmorillonite) composition. The absorbed complexes of clay minerals, as in the shallow-water deposits, are dominated by bivalent ions of calcium. At certain periods, the clay material settles to the bottom together with a great amount of fine humus organic matter, whose accumulation causes the formation of gas-prone sediments similar to the black shales from the Atlantic margins [192–194].

4.3.3.   MAXIMUM OF TRANSGRESSION

During the transgressive stage of sedimentation basin development, deposition of clay material occurs mainly within the deep-water part of shelf and in the adjacent sea basin at a considerable distance from the shoreline, where the supply of sand-silt material is practically absent. The area of deposition is characterized by hydrodynamically quiet regime of bottom waters and by the accumulation of sediments with horizontal bedding, laterally persistent, containing a specific deep-water fauna complex. The depths of sediment formation vary from 200 to 500 m (Fig. 4.5, basin evolution stage III).

At this stage of basin evolution, clay material is supplied from the deepest horizons of weathering crust profiles within source areas, where predominantly distributed minerals are montmorillonites and mixed-layer illite-montmorillonite varieties. In accordance with the conditions of mechanical sedimentary differentiation, the particles of montmorillonite and mixed-layer minerals, as the least in size, are brought to the farthest from shore basinal areas with greatest water depths. Here, under the conditions of hydrodynamically quiet regime of bottom waters, they sink to the bottom of the basin. In the absorbed complex of settling minerals, univalent (mainly Na^+) cations prevail. This creates conditions for the formation of homogeneous fine grained sediments, which later become reliable screens.

The above-considered model of seal formation is in agreement with the concepts of a number of authors [99, 195]. According to their data, in the majority of oil- and gas-bearing basins, the most finely elutriated and thickest horizons of clays with high contents of montmorillonite minerals occur in the upper, maximally transgressive, parts of sections. These deposits correspond to the deep-water shelf facies of marine sediments, which complete the sedimentary cycle (Fig. 4.6). The middle part of the section is composed of clays with admixture of fine silt material and mineral components of mixed-layer illite-montmorillonite and illite composition, accumulated in the medium-depth environments. The lower part of the section is composed of silty clays and silts, mainly of chlorite and illite composition, formed in the near-shore agitated shallow-water environment and correspond to the beginning of transgressive stage in the sedimentation basin evolution.

Summing up the aforesaid, one may draw a conclusion that initial environments of clay sediment deposition determine the following:

1. Grain-size composition of the sediment – the proportion of sand-silt and clay fractions. The formation of grain-size composition is controlled by the hydrodynamic regime of bottom waters in a sedimentation basin, on which the conditions of mechanical differentiation of sediment material depend. Clay sediments with a high content of sand-silt-sized particles form in the shallow-water coastal zone and in the shallow-water part of shelf nearer to river deltas,

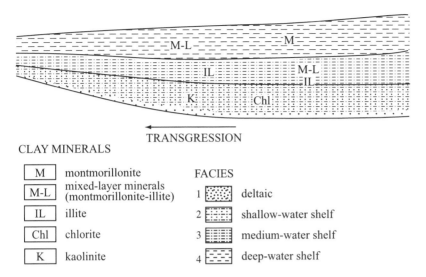

CLAY MINERALS

M	montmorillonite
M-L	mixed-layer minerals (montmorillonite-illite)
IL	illite
Chl	chlorite
K	kaolinite

FACIES

1 deltaic
2 shallow-water shelf
3 medium-water shelf
4 deep-water shelf

Fig. 4.6. Distribution of clay minerals in an oil- and gas-bearing basin during its evolution (in a cross-section).

more fine grained clays with insignificant contents of sand-silt material settle down in more distant and deeper-water parts of a sedimentation basin.

2. Mineral composition of clay sediments determined by the climatic conditions and character of tectonic activity within the source area, as well as by the mineral differentiation occurring in a sedimentation basin during sedimentogenesis.

3. The composition and contents of the absorbed complex of Na^+, K^+, Mg^{2+}, Ca^{2+} cations, the Na^+/Ca^{2+} proportion in the absorbed complex, and the absorption capacity, all depending on the facies-geochemical conditions of sediment accumulation. In the coastal near-delta zone, clay minerals contain a considerable quantity of bivalent calcium cations promoting the formation of large aggregates of clay particles, which leaves its imprint on the primary structure of clay sediment. In deeper waters, clay particles contain mainly univalent sodium cations, which cause the dispersion of the clay component of sediments. These sediments have a fine grained structure and high absorption capacity [107].

4.4. CLASSIFICATION OF CLAY SEALS ACCORDING TO THEIR DEPOSITIONAL ENVIRONMENTS AND SCREENING PROPERTIES

A number of classifications of clay seals for oil and gas fields are known at present. All of them have been worked out for some specific region and based on the subdivision of seals or according to one or several characteristic features. The classifications developed by various authors for West Siberia became the most widely used. Thus, B.P. Filippov [196], taking as a basis the Upper-Jurassic and Lower-Cretaceous clay sequences that screen gas accumulations in the Berezovo region, has distinguished the true seals and the semi-seals differing in their permeability. The true seals have a homogeneous mineral composition and absorption capacity exceeding 30 mg-equiv. per 100 g of rock, their characteristic feature is the sharp predominance of exchangeable sodium over calcium. The semi-seals are characterized by the variegated composition of clay mineral components and by the predominance of exchangeable calcium.

T.F. Antonova [197] has worked out a classification of clay seals for hydrocarbon fields in the central part of the West-Siberian lowland. The classification contains a schematic description of the mineral composition of Jurassic clay sequences and data on the density of clays, content of a sand-silt fraction in them, proportion of sodium and potassium cations in the absorbed complex.

G.Eh. Prozorovich [107] has worked out a classification of clay seals taking into account their diffusion-, filtration-, and fracture-related permeability. He defines a clay seal for oil and gas as a sequence of predominantly low-permeability rocks of different composition and genesis, whose diffusion-, filtration-, and fracture-related permeability within the highest hypsometric parts of hydrocarbon traps is low enough. This creates the conditions, under which this sequence detains a considerable part of hydrocarbons in the reservoir rock overlain by it. As a basis for the seal classification, G.Eh. Prozorovich [107] has used the degree of lithologic heterogeneity of seals and the character of their distribution within the West-Siberian oil- and gas-bearing basin. Proceeding from these features, he has differentiated homogeneous and heterogeneous shale and sandy-silty-clay seals characterized by different ranges of the aggregate sandstone and siltstone thickness in percentage of the total seal thickness. The indicated parameter depends on the number and thickness of sandstone and siltstone interlayers and is very informative in predicting the clay seals in the regions poorly studied by drilling. In addition to this, the classification by G.Eh. Prozorovich [107] subdivides the clay seals depending on their distribution area. Within the West-Siberian basin, the regional, zonal, and local seals are distinguished. Their description includes data on the composition, epigenetic transformation of clay minerals in different zones of lithogenesis, as well as data on their density, porosity, and other properties.

The above-considered classifications are based on certain characteristics of clay sequences and do not take into account the general regularities in their lithologic-genetic alterations associated with the facies (depositional environment) control. Meanwhile, depositional environments are the most important criterion for the formation of certain types of seals and may be considered as a fundamental index in their classification. Indeed, the initial depositional environments of clay sediments determine, as has been noted above, their lithologic composition, homogeneity, and thickness, which are responsible for their characteristic structural features at the facies level. A factor closely related to the depositional environment is the geochemical environment of sediment accumulation, on which the composition and content of the absorbed cation complex, the authigenic mineral formation, the composition and quantity of organic matter depend. The primary facies features exert a considerable influence on the character of sediment transformation in the course of lithogenesis, as they control the specificity of change with depth in their mineral composition, porosity, and permeability, as well as the development of microfractures. All this evidences that the facies features of clay sequences, along with the specificity of clay microstructure at the rock level, are the critical factor in estimating the clay screens.

Proceeding from the above, the present authors have worked out a classification of clay seals for oil and gas fields according to their facies characteristic, which encompasses clay sediments from the deep-water facies to the sediments of coastal highly agitated shallow-water environment (Table 4.1).

It follows from the classification, that clay sediments deposited in the environment of deep-water shelf and adjacent part of sea at depths of 300–500 m, under the conditions of a quiet hydrodynamic regime are characterized by the homogeneous structure, predominance of montmorillonite in the composition of clay minerals with particle size $<0.1 \mu m$, content of silt material up to 10%, carbonate content of more than 1%, organic matter of the sapropel type. The sequence thicknesses are from few meters to tens of meters. The prevailing cations in the exchange complex are those of Na^+, the Na^+/Ca^{2+} ratio being 6–12. The ability of these clays to form microfractures in the process of lithogenesis is low at depths shallower than 5000 m. The permeability of seals formed by them is 10^{-5} mD to gas, the diffusion coefficient is $10^{-10}-10^{-9} cm^2/sec$. The clay sediments under consideration form seals of class I, which serve as reliable screens for oil and gas fields (Table. 4.1).

Sediments of the deep-water shelf environment accumulate at depths of 200–300 m under the conditions of quiet hydrodynamic regime and form homogeneous sequences up to some tens of meters thick. The dominant clay minerals are montmorillonite and mixed-layer (illite-montmorillonite) varieties with particle sizes of 0.1–0.2 μm. The content of sand-silt material does not exceed 10–20%, that of carbonates is more than 1%, organic matter is represented by the sapropel type. Microfracturing develops weakly in the process of lithification. Absolute permeability to gas is 10^{-5} mD, diffusion coefficient is $10^{-10}-10^{-9}$ cm^2/sec. When lithified, the clay sediments of deep-water shelf environment form clay seals of class II, impermeable for oil and gas (Table 4.1).

Sediments of the middle-shelf environment accumulate at depths of up to 200 m under the conditions of a relatively quiet hydrodynamic regime and form largely homogeneous clay sequences up to several tens of meters thick. The clay fraction of the sediments is dominated by mixed-layer (illite-montmorillonite) varieties and illite with particles no less than 0.3 μm in size. The content of sand-silt material is 20–30%, the content of carbonates is 1–2%; sapropel varieties prevail among organic remains. The ratio of exchangeable Na^+ and Ca^{2+} is 3–5. In lithification, the clays show weak microfracturing and are characterized by the following screening properties: absolute permeability to gas – 10^{-4} mD, diffusion coefficient – $10^{-8}-10^{-7}$ cm^2/sec, which makes them impermeable to oil and diffusionally low permeable to gas. According to their screening properties, they rank with the class III seals (Table 4.1).

Sediments of the peripheral part of delta area accumulate at depths of up to 100 m under the conditions of weak turbidite current influence. They form a sequence of clay sediments interbedding with silt, with total thickness amounting to several tens of meters. The clay fraction of the sediments is mainly represented by illite and mixed-layer (illite-montmorillonite) varieties with particle sizes ranging from 0.3 to 1 μm. The content of sand-silt material is 30–40%, the content of carbonates is 2–3%; there are admixtures of organic matter of the sapropel-humus type. The ratio of exchangeable Na^+ and Ca^{2+} is 2–4. The clays are capable of moderate microfracturing in the process of lithogenesis. The absolute permeability of seals formed by them is 10^{-3} mD to gas, the diffusion coefficient is 10^{-7}–10^{-6} cm^2/sec, which forms a basis for ranking them as class IV seals – diffusionally permeable to gas and impermeable to oil (Table 4.1).

Sediments of the shallow shelf environment accumulate at depths of up to 50–70 m under the conditions of relatively active water dynamics and form heterogeneous sequences of interbedding clay, silt, and sand material up to several tens of meters thick. The clay layers have a polymineral composition, the clay fraction is dominated by non-swelling minerals (illite, kaolinite, and chlorite) with particle sizes of 0.5–3 μm. The content of the sand-silt fraction is 40–50%, the content of carbonates is up to 5%, organic matter of the humus type with an admixture of the sapropel type is present. The ratio of exchange cations Na^+/Ca^{2+} is 1–3. In the course of lithogenesis, the clays may develop microfracturing and form seals with the following screening properties: absolute permeability to gas – 10^{-2} mD, diffusion coefficient – 10^{-6}–10^{-5} cm^2/sec. The seals formed by them are permeable to gas and low permeable to oil, which allows their ranking with class V (Table 4.1).

Sediments of the near-shore agitated shallow-water environment form at depths of up to 30–50 m under the conditions of active water dynamics. They compose heterogeneous sedimentary sequences up to several tens of meters thick made up of silty clays with silt and sand interlayers. The clay layers contain up to 50–70% of sand-silt material, up to 5% of carbonates, and an insignificant admixture of humus-type organic matter. The clay fraction is polymineral in composition with the predominance of non-swelling minerals (kaolinite, chlorite, and illite) having particle sizes of 0.5–3 μm. The ratio of exchangeable Na^+ and Ca^{2+} does not exceed 1–3. The absolute permeability of the sequence in a lithified state is 10^{-2} mD to gas, the diffusion coefficient is slightly higher than 10^{-5} cm^2/sec, a characteristic feature is considerable microfracturing. The seals formed of the shallow-shelf sediments do not screen hydrocarbons individually and rank according to their screening properties with class VI (Table 4.1).

Sediments of the near-shore highly agitated shallow-water environment form at depths of 20–30 m under the conditions of active hydrodynamics. The sequences formed of them in lithification are heterogeneous in structure and consist of clayey silts with silt and sand layers. The layers of clayey silts contain up to 70–80% of sand-silt material, up to 5% of carbonate material and an admixture of humus-type organic matter. Dominant clay minerals are kaolinite, chlorite, and illite with particle sizes of 0.5–10 μm. The ratio of exchangeable

Na^+ and Ca^{2+} does not exceed 1. In lithification, these deposits are susceptible to microfracturing. Absolute permeability of the sequences to gas is $<10^{-1}$ mD, diffusion coefficient is slightly higher than 10^{-5} cm^2/sec. The clay sequences formed in the near-shore highly agitated shallow-water environment are not able to screen oil and gas independently. According to their screening properties, they rank with class VII seals (Table 4.1).

As is seen from the given classification, the screening properties of seals change in a regular manner depending on the depositional environments controlling their formation. The latter affect both the composition of clay material and the structure of clay sequence, including the presence of well permeable silt-sand layers in it. The most homogeneous in lithologic composition seals of classes I and II form in the zones of basins least of all subjected to hydrodynamic effects. Their characteristic features are the minimal contents of sand-silt material, predominance of fine grained clay minerals having a swelling crystal lattice, and high content of exchangeable sodium in the absorbed complex. The seals of classes I and II form a group of seals providing good screens to accumulations of both oil and gas.

With the passage to the middle-shelf and peripheral part of delta environments, the sediments become less homogeneous: the sedimentary sequences that form here show an increase in the content of sand-silt layers, the non-swelling mineral illite, along with the mixed-layer varieties (of the illite-montmorillonite type), becomes common among the clay minerals. The clay seals of classes III and IV developed from these sediments in the process of lithogenesis form a group of average-quality seals, impermeable to oil and diffusionally permeable to gas.

The number and thickness of permeable silt-sand layers increases significantly in sediments of the shallow-shelf and shallow-water coastal environments, where the most heterogeneous and sand-enriched sediments accumulate. The clay layers are characterized, along with the increase in sand-silt fraction, by the dominance of non-swelling minerals – illite, kaolinite, and chlorite. The clay seals of classes V–VII developed under these conditions form a group of seals with low screening properties, unable to provide reliable screens for hydrocarbon accumulations.

So, the facies principle may be a reliable and universal basis for the classification of clay seals and prediction of their screening properties.

5

Facies Analysis of Clay Seals in Oil- and Gas-Bearing Basins

5.1. THE PRINCIPAL CRITERIA FOR FACIES ANALYSIS

As has been shown in the previous chapter, the screening properties of clay seals are determined by their depositional environments reflecting the stage of sedimentogenesis. Reconstruction of the initial conditions controlling sediment accumulation makes it possible to examine the course of sedimentogenesis and geological history of oil- and gas-bearing basin, as its sediments bear evidence of the principal regional-scale events.

The conditions and specificity of sediment accumulation are reflected in primary genetic features of a sediment which becomes transformed into a rock at the following stages of sedimentary basin evolution.

So, clay seals have both the primary genetic features reflecting their genesis and the secondary features characterizing their subsequent history.

To establish the facies type of clay seals, a complex methodological approach must be applied based on the analysis and synthesis of primary genetic features of sediments, which are reflected in the rock and characterize its depositional environment. This is the method of lithological-facies analysis using the principle of actuality and worked out for sedimentary formations by Y.A. Zhemchuzhnikov [198], P.P. Timofeev [199], and others. The basis of the method is the recognition in a rock of its genetic features reflecting its origin, the principal of which are the following:

1. Granulometric composition of rock – grain size distribution of sedimentary material reflecting the process of mechanical differentiation of sediment.
2. Textural features of rock (type of bedding), which characterize the depositional environment dynamics.
3. The capacity and composition of the absorbed complex, reflecting the depositional environment geochemistry.
4. The composition of fauna – shallow-water or deep-water.
5. The presence or absence of organic matter admixture, including terrigenous organic matter, evidencing for the basin depth and distance to the source area.
6. The contents of mineral components, including light and heavy minerals, reflecting the character of source areas.

7. The contents of carbonate material and the presence of authigenous minerals, characterizing the medium of sediment deposition.

The genetic nature of sediments is imprinted in the material composition of rocks, which is clearly revealed in detailed microscopic, X-ray, and chemical examination.

The aggregate of all the features allows a judgment on the sediment nature and depositional environment dynamics, as well as its comparison with sediment accumulation in present environments – zones of deltas, coastal and deep-water marine areas.

Reconstruction of the nature of each individual bed, revealing of regularities in the distribution of beds throughout the section and over the area allow one to reconstruct the evolution of the sediment accumulation process during a certain age interval.

An elementary paleogeographic unit is a facies. The term «facies» means not only a complex of physical-geographical conditions of a depositional environment in which one or several kindred genetic types of sediments have been formed, but also the sediments themselves having the corresponding complex of primary genetic features [199].

The whole variety of clay seals differing in their screening properties can be reduced to the following three principal facies complexes of sediments:

1. *The facies of clay sediments deposited in deep-water shelf and adjacent marine basin environments – the seals of classes I and II.*

The sediments are characterized by the following genetic features: high degree of sedimentary material dispersion and negligible admixture of silt-size particles; horizontal bedding reflecting quiet sedimentation conditions in the absence of deep-water fauna; admixture of sapropel organic matter of the algal type; presence of authigenous dolomites; low contents of heavy- and light-fraction minerals. The prevailing mineral components of clays are montmorillonite and mixed-layer (illite-montmorillonite) varieties (see Table 4.1).

2. *The facies of clay and clay-silt sediments deposited in the environments of medium-depth shelf and peripheral part of submarine delta – the seals of classes III and IV.*

The following chief genetic features are characteristic of these sediments: the clays are low-silty; among mineral components, along with montmorillonite and mixed-layer (montmorillonite-illite) varieties, also illite is present in considerable quantities. The bedding is horizontal and horizontal-undulating, locally thin, cross striated, flattening and discontinuous; rare roiling and slumping structures are present. Other characteristic features are the presence of algal-type sapropel and humus-type vegetable organic matter, as well as of authigenous dolomites, a noticeable admixture of heavy- and light-fraction minerals.

3. *The facies of clay-silt and silt sediments deposited in the shallow-water shelf and coastal agitated shallow-water environments – the seals of classes V, VI, and VII.*

The following chief genetic features are characteristic: the clays contain a significant admixture of silt-sand material (silty clays and clayey silts); the prevailing mineral components are kaolinite, chlorite, illite; cross-bedded structures (unidirectional and not unidirectional cross bedding, small-scale cross striation, and discontinuous bedding), roiling and slumping structures are common. The structures reflect the active hydrodynamics and high rates of sediment accumulation. The prevalent organic matter is vegetable humus with fragments of higher plants' tissues. Fragments of shallow-water fauna and authigenous dolomites are present. A characteristic feature is the high contents of heavy- and light-fraction minerals.

So, the formation of seals with different screening properties is related to the depositional environments controlling their origin. The fine grained clays with high screening properties (seals of classes I–II) were deposited in the deep-water shelf environment, under quiet hydrodynamic regime; the seals with average screening properties (of classes III–IV) were formed nearer to the shore under more active hydrodynamic conditions; the seals with low screening properties (of classes V, VI, and VII) were formed in the coastal shallow-water environment affected by various flows and currents.

5.2. FACIES ANALYSIS OF CLAY SEALS IN SEDIMENTARY BASINS

5.2.1. WEST-SIBERIAN BASIN

The basin contains the largest oil and gas fields of Russia: Surgut, Nizhnevartov, Urengoy, Taz, Zapolyarnoe, Us't-Balyk [173, 200].

The stratigraphic section of the West-Siberian basin is composed of largely sand-silt-clay sediments of Mesozoic age overlain by Paleogene deposits (Fig. 5.1). The Jurassic sediments (Tyumen', Vasyugan, Georgiev, and Bazhenov formations) are composed of clays belonging to the marine basinal and deep-water shelf facies: these are fine grained, horizontally bedded clays with marine deep-water fauna. They alternate with the silt-clay and silt sediments with lenticular, cross-undulated, and cross bedding belonging to the shallow-water shelf and coastal facies. This indicates the constant change of transgressions and regressions.

Thus, in Callovean and Oxfordian time (Vasyugan formation), sea covered the most part of West Siberia, was relatively deep and had normal salinity: dark-gray, horizontally bedded clays with marine fauna accumulated in it. In the late Oxfordian, there was an intense supply of clastic material, shallow-water environments became prevalent, and silty cross-bedded sediments accumulated. The formation thickness is 50–90 m. In Kimmeridgian time (Georgiev formation), the West-Siberian marine basin expands again to accumulate thin horizontally bedded clays. The sediments are 2–15 m thick. In Volgian time (Bazhenov formation), the most part of the territory represented a deep-water shelf. Water was of normal salinity. The accumulated sediments were horizontally bedded, finely elutriated black clays enriched in organic matter. The sediment thickness is 10–30 m. The Bazhenov formation clays occurring at depths of 2700–2800 m serve as seals of classes I and II for oil fields of the Middle-Ob' region (Us't-Balyk, Samotlor). At places, the presence of low-silt clays forming the

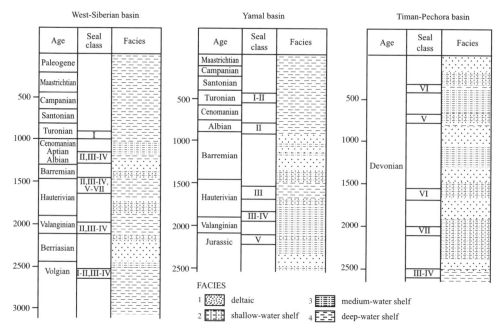

Fig. 5.1. Stratigraphic and facies sections and the position of clay seals of different classes within the West-Siberian, Bureya, South Yamal, and Timan-Pechora basins.

III–IV class seals is observed at the indicated depth evidencing for the local occurrence of middle-shelf environments.

The Cretaceous Berriasian-Valanginian sediments (Megion, Culomzin, Tar formations) are composed of alternating clays, siltstones, and sandstones of marine and coastal genesis. The sediments are 630–900 m thick. In the first half of the Berriasian, thin horizontally bedded clays accumulated in the deep-water marine environment, then the supply of clastic material from source areas increased resulting in the shallowing of the basin and accumulation of sand-silt deposits. At the end of Berriasian and in Valanginian time, the sedimentation basin deepened again and marine deep-water environments with normal water salinity and accumulation of fine grained clays were established over the most part of the territory. These clays have formed the II class seals (Cheuskin unit). In some zones of the basin, in the medium-depth shelf environment, low-silt clay sediments accumulated to form seals of classes III–IV. The seals occur at depths of 1900–2000 m in the Nizhnevartov oil-producing region.

The upper part of the Valanginian, the Hauterivian and Barremian (Vartov formation) form a sequence 350–600 m thick. The sedimentary section represents an alternation of sandstones, siltstones, and green clays of continental genesis. Some silt-clay layers having gray color formed in the environments of coastal shallow waters and peripheral parts of deltas.

This stage of sediment accumulation was characterized by further shallowing of the sedimentation basin and reduction of deep-water marine areas. The sediments accumulated within alluvial plain, coastal shallow-water areas, central and peripheral parts of deltas and shelf. Concurrently, organic matter brought from the continent accumulated there, which created favorable conditions for the formation of oil- and gas-bearing sequences. The formation of clay seals occurred under various conditions: the fine grained clays accumulated in the deep-water shelf environment formed seals of class II, the shelf clays accumulated in the medium-depth environments formed seals of classes III–IV, and the siltstones of the coastal shallow-water environment formed seals of classes V–VIII. In the Surgut area, the seals of classes II, III, IV, V, VI, and VII occur at depths of 1500–1600 m and overlie the oil-bearing near-shore marine sandstones.

The Aptian deposits (Alym formation) are 30–90 m thick. In the lower part of the section, dark-gray clays of deep-water shelf occur followed upwards by black clays with limestone intercalations representing the marine deep-water facies. The top of the section is composed of shelf clays with thin interlayers of siltstones and sandstones. There was an extensive transgression in the Aptian, with sea penetrating deep into the continent and with clay sediment accumulation; sandy-silty-clay sediments were deposited over significant areas in the shallow-water environment. At the end of Aptian time, a stagnation regime was established leading to the enrichment of sediments with bitumen. The deep-water Aptian clays have formed seals of class II, the medium-depth shelf clays – seals of classes III–IV. Within the Nizhnevartov field they occur at depths of 1100–1200 m.

The upper part of the Aptian-Albian-Cenomanian section (Pokur formation) is 200–800 m thick. The lower part of the sequence is composed of interbedding sandstones, siltstones, and clays containing abundant carbonaceous detritus. In the upper part, there are cross-bedded sandstones with rare interlayers of siltstones and silty clays. In the Albian, alluvial plain environments dominated over the larger part of West Siberia. In Middle-Albian time, a new transgression followed and delta fans of large rivers falling into the sea basin became widely developed.

The Turonian (Kuznetsov formation) is 20–40 m thick. The section is composed of dark-gray, almost black clays. In the Turonian, marine transgression intensified and deep-water marine environments became dominant over the larger part of the territory till the end of Cretaceous time. The highly dispersed clays accumulated in the deep-water environments, which serve as seals of class I for the Urengoy, Taz, Zapolyarnoye gas fields and occur at depths of 800–1000 m.

The cyclic character of the West-Siberian basin evolution during its Jurassic and Cretaceous history [107, 195] has manifested itself in the five-fold recurrence of deep-water sedimentation conditions – termination of regressive-transgressive cycles and gradual intensification of transgression.

The maximum-transgression conditions have manifested themselves in the accumulation of fine grained clays of marine deep-water environments and resulted in the formation of I–II class seals. During the Mesozoic history, these conditions were manifested five times – in Volgian time, in the Valanginian, Hauterivian, Aptian, and Turonian. The Upper-Jurassic clays occurring at depths of 2700–2800 m serve as seals for oil fields of the Middle-Ob' region: Ust'-Balyk, Samotlor. The Aptian clays occurring at 1100–1200 m serve as seals for the Nizhnevartov oil field, the Turonian clays (depth 800–1000 m) screen gas pools in the Urengoy, Taz, Zapolyarnoe fields (Fig. 5.1).

The middle of transgression has manifested itself by the accumulation of clay and low-silt clay sediments of the medium-depth shelf and peripheral part of delta environments. These are clay seals of classes III–IV. The indicated sedimentation conditions were repeated in West Siberia four times during the Mesozoic sedimentation basin development: in Volgian time, in the Valanginian, Hauterivian, and Albian. They are registered in the Shaim area within the Volgian stage at a depth of 2800–2900 m, in the Nizhnevartov area – in the Valanginian at depths of 2100–2300 m; in the Surgut area – in the Hauterivian at depths of 1500–1600 m, and in the Aptian–Albian – in the Nizhnevartov area at depths of 1100–1200 m (Fig. 5.1).

The beginning of transgression is characterized by the presence of clay layers of small thickness with an admixture of sand-silt material, cross-bedded, alternating with silt and sand interlayers and formed in the zone of coastal marine shallow-water sedimentation and shallow-water shelf. These clays form seals of classes V, VI, and VII. In the West-Siberian basin, the seals of these classes are Hauterivian in age and occur at depths of 1500–1600 m.

Figure 5.2 shows a model for the West-Siberian basin development in Cretaceous time [195]. One can see there the area of dominant deep-water environment and accumulation of clays forming the seals of classes I and II;

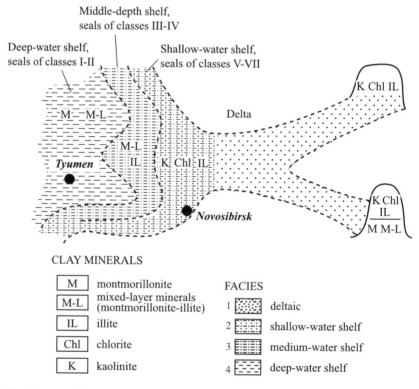

Fig. 5.2. A model for the West-Siberian basin development and a plan view of clay seal distribution in Mesozoic time.

the deep-water shelf gives place to the medium-depth shelf, where silty clays forming the seals of classes III–IV accumulated, and farther to the east it passes into the coastal shallow-water zone with accumulated silts – the seals of classes V, VI, and VII. The coastal shallow-water area joined to a river delta supplying clay material from the Sayan region and Middle Siberia.

5.2.2.   YAMAL BASIN

The basin section is composed of Mesozoic sediments 2800 m thick [201] (Fig. 5.1). As to the conditions of its origin, the Yamal basin is much similar to the West-Siberian basin.

The base of its section is composed of Lower to Middle Jurassic sediments (analogues of the Tyumen' formation) – continental and coastal marine sands-silts-clays with coal interlayers, 560 m thick. The clay layers with an admixture of silt material form a seal of class V occurring at a depth of 2100–2200 m. Higher in the section, there are Upper Jurassic-Hauterivian deposits; the shelf bituminous silty argillites encountered in the lower part of this sequence form a III–IV class seal occurring at depths of 1800–1900 m. Overlying them are sand-silt deposits of coastal shallow-water environment, they are followed by the medium-depth shelf clays, which also form seals of class III at depths of 1600–1700 m. The sequence is 500 m thick. In the Barremian-Cenomanian time, a sequence 1200 m thick has formed. The sand-silt deposits of this age accumulated in the coastal shallow-water environment; however, the following Albian marine transgression resulted in the accumulation of fine grained clays in the deep-water shelf environment. This created conditions for the formation of class II seals occurring at depths of 800–900 m.

The section is topped by Turonian-Maastrichtian clay deposits (Kuznetsov, Berezov, Gan'kin formations) of the deep-sea environment represented by fine grained clays – the seals of classes I and II occurring at depths of 400–500 m.

Consequently, it was 5 times in the Yamal basin history that conditions for the formation of clay seals were created. Once (at a depth of 2100–2200 m), the class V seals have formed here, twice (1800–1900 m and 1600–1700 m) – the seals of classes III–IV, and twice (800–900 and 400–500 m) the seals of classes I and II have formed (Fig. 5.1).

5.2.3.   TIMAN-PECHORA BASIN

The Timan-Pechora basin is composed of Devonian sediments represented mainly by sandstones and siltstones with thin interbeds of silty clays [202] (Fig. 5.1). The sediments are 2800 thick.

During the Devonian, there was a marine transgression with submergence of the whole sedimentation basin. This resulted in the establishment of medium-depth shelf conditions in Early Devonian time and in the accumulation of silty clays that formed seals of classes III–IV at depths of 2500–2600 m. Later, in the Middle Devonian, the coastal shallow-water environments recurred four times resulting in the accumulation of silty clays serving as seals of classes V–VII.

At depths of 2000–2100 m, the basin contains a clay seal of class VII, at depths of 1500–1600 m – a seal of class VI, at a depth of 700–800 m – a seal of class V, and at a depth of 400–500 m – a seal of class VI.

5.2.4.   VOLGA-URAL BASIN

The basin section is composed of sandy-silt, clay, and carbonate deposits of Devonian, Carboniferous, and Permian ages, having continental, near-shore marine, and marine genesis [202]. The sediments are 3000 m thick (Fig. 5.3). In the Devonian, the continental environments changed over to the shallow-water and coastal marine environments, which resulted in the accumulation of silty clays – the seals of class V occurring at depths of 2700–2800 and 2300–2400 m. In the Carboniferous, due to the extension of transgression, the middle-shelf

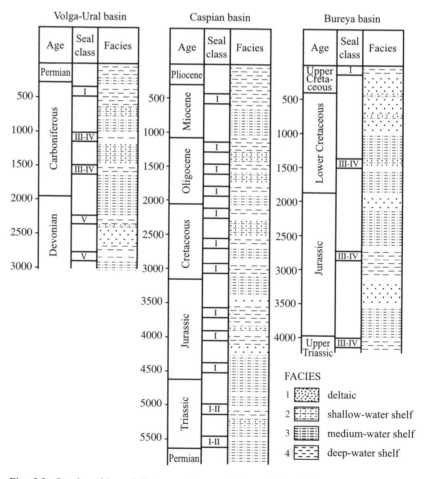

Fig. 5.3. Stratigraphic and facies sections and the position of clay seals of different classes within the Volga-Ural, Caspian, and Bureya basins.

conditions occurred twice with formation of low-silt clays – the seals of classes III and IV occurring at depths of 1500–1600 and 1100–1200 m. At the end of the Permian – in the Carboniferous, marine transgression resulted in the establishment of deep marine conditions with accumulation of fine grained clays serving as seals of class I and occurring at depths of 400–500 m.

5.2.5. CASPIAN BASIN

The basin section encompasses a stratigraphic interval from Permian through Pleistocene, the sediment thickness being 6500 m [203] (Fig. 5.3). It is composed of sand-silt-clay sediments of marine and coastal marine genesis. The section shows a successive alternation of coastal marine shallow-water environments manifested in the accumulation of sand-silt cross-bedded deposits with deep-sea environments expressed in the formation of fine grained clays. The alternation of the indicated facies reflects the transgression and regression stages of the sea basin.

In the Permian period, fine grained clays accumulated to form the seals of class I occurring at a depth of 6100 m; in the Triassic, two transgressions resulted in the accumulation of two units of deep-water fine grained clays – the seals of classes I and II occurring at depths of 5600 m and 5000 m. In Jurassic time, three transgressions resulted in the accumulation of fine grained clay units – the seals of class I at depths of 4500 m, 4000 m, and 3700 m. In the Cretaceous, the periods of maximal transgressions are responsible for the accumulation of clays forming seals of class I at depths of 2900 m, 2500 m, and 2300 m. In the Oligocene time, three maximal transgressions favored the accumulation of clays – the seals of class I at depths of 1700 m, 1500 m, and 1200 m. The Miocene transgression was expressed in the accumulation of clays – the seals of class I occurring at a depth of 500 m.

5.2.6. BUREYA BASIN

The Bureya basin is composed of Mesozoic (Triassic, Jurassic, Cretaceous) sediments of marine, near-shore marine, and continental genesis, 4000–6000 m thick [204] (Fig. 5.3). During the Mesozoic history, the continental conditions of sediment accumulation were repeatedly interrupted by marine transgressions resulting in the formation of sand-silt-clay sediments in the marine coastal shallow-water and medium-depth shelf environments, as well as of marine sediments in the deep-sea environments.

Thus, the Upper Triassic transgression has led to the development of a medium-depth shelf and accumulation in this environment of low-silt clays forming seals of classes III–IV at depths of 4000–4100 m. Later, the continental and coastal marine depositional environments set in with the accumulation of sand-silt sediments. A transgression in the middle of the Jurassic stage of basin development resulted in the establishment of medium-depth shelf environment and accumulation of low-silt clays that formed the III–IV class seals at depths of 2700–2800 m. The Lower Cretaceous transgression repeated the medium-depth shelf conditions and accumulation of silty clays – the seals of classes III–IV occurring at depths of 1400–1500 m.

In the Upper Cretaceous, the maximum transgression occurred expressed in the accumulation of fine grained horizontally bedded clays forming the seals of class I at a depth of 100 m.

Thus, in the Bureya basin, conditions for the formation of III–IV class seals (at depths of 4000–4100, 2700–2800, and 1400–1500 m) were created three times and for the formation of seals of class I (at a depth of 100 m) – once.

6

The Composition of the Facies Types
of Clay Seals and Its Change with Depth

6.1. THE COMPOSITION OF THE FACIES TYPES OF CLAY SEALS

As has been said in Chapter 4, the presence of seals with different screening properties is connected with depositional environments controlling the accumulation of clay material. Sediments formed within different zones – coastal, shallow-water, and deep-water – of a sedimentation basin differ in their composition. These differences are chiefly determined by: (1) the composition of clay minerals and their quantitative proportions, (2) the admixture of sand and silt material, (3) the absorption capacity and the absorbed complex composition, (4) the admixture of organic matter of terrestrial type [105–107, 205–214].

6.1.1. FACIES OF THE DEEP-WATER SHELF AND ADJACENT SEA BASIN

Clay sediments of deep-water shelf have zonal–regional distribution and significant variations of thickness: from few to several tens of meters and more. The forming sequence of clay sediments is, as a rule, uniform, with obscure horizontal bedding reflecting sedimentation under the conditions of hydrodynamically quiet regime of bottom waters.

A distinctive characteristic of deep-water sediments is their high dispersion with predominance of particles <0.005 mm, whose contents is commonly no less than 80–90%. The admixture of fine (0.01–0.005 mm) silt material is quite insignificant and seldom amounts to 20%. The large-size silt and sand particles are extremely rare. Silt material is encountered in the form of individual inclusions submerged into a fine grained clay mass.

The mineral composition of deep-water clay sediments is represented by montmorillonite and illite-montmorillonite mixed-layer varieties with the contents of swelling layers from 40–60% to 60–80%. The non-swelling minerals – kaolinite, illite, chlorite – are contained as admixtures. The formation of the mineral composition of sediments is closely associated with the conditions of mechanical sedimentary differentiation, when particles <0.1 μm in size characteristic of montmorillonite and those 0.1–0.2 μm in size characteristic of mixed-layer minerals of illite-montmorillonite composition settle down under the hydrodynamically quiet conditions.

The carbonate material makes up no more than 1–2% and is represented by the remains of foraminiferas, cocolithophorids, mollusks, confirming the sedimentation at depths of 200–500 m. The presence of organic matter of the sapropel-algal type is registered; the terrestrial humus-type plant organic matter is practically absent.

The sorption capacity and the composition of absorbed cations are in complete agreement with the granulometric and mineral composition of sediments. In the clays of marine deep-water origin, exchangeable sodium is largely dominant over calcium. According to G.Eh. Prozorovich [107], the content of exchangeable sodium makes up no less than 60% of the sum of exchange cations. The exchange capacity is no less than 35–40 mg-equiv./100 g.

The deposited clay sediments are the thermodynamically unbalanced open systems, whose composition undergoes significant transformations in the course of lithogenesis. The most important factor of such transformations is the change in geochemical and pressure–temperature conditions with sediment burial. As the properties of clays are closely associated with their composition, the observed changes are of great importance in studying the regularities in the formation of the screening properties of seals [106, 107, 210, 211, 215, 216].

The lithogenetic transformation of sediments of the deep-water shelf and basinal facies results in the formation of clay seals of classes I and II, which are characterized by the highest screening properties.

A study of the granulometric composition of this type of seals occurring at different depths has made it possible to examine the change with depth in the ratio of sand-silt (aleurite) fractions (sa) to the clay fraction (c). As is seen from Figure 6.1, the quantity of silt and sand particles in the initial sediment is extremely low and is characterized by the ratio value $K_{sa/c} = 0.05$. As the sediments subside, this ratio grows in a regular way and is about 0.14 for a depth of 2.5 km. The obtained correlation between the ratio $K_{sa/c}$ and the depth of clay occurrence is as follows:

$$K_{sa/c} = -1.7073 + 0.41317 \times \ln h; \quad R = 0.84 \tag{6.1}$$

where R – correlation coefficient.

The above data evidence that some reduction of the material dispersion with depth as compared to its initial composition is observed in the seals of types I and II. Such a phenomenon may be related to the formation of larger secondary minerals in the process of recrystallization, the compaction and cementation of clay mineral aggregates with the formation of platy polycrystalline overgrowths. An important factor of sediment differentiation depending on depth may be the evolution of source areas and sedimentation basin causing the supply of coarser material to the deep-water part of the basin at the early stage of its evolution (the lower part of the section) in comparison with the later stages (the upper part of the section).

Despite the development of processes tending to reduce the dispersion of I and II type seals with depth, these sediments continue to be extremely high-dispersion systems even at considerable depths: the content of clay fraction in

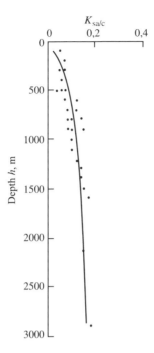

Fig. 6.1. Change in the ratio of sand-silt and clay fraction contents ($K_{sa/c}$ = (sand + silt)/clay) with depth for the seals of classes I and II.

them at depths of 2–3 km remains 6–7 times higher than the quantity of sand- and silt-size particles (Fig. 6.1).

A study of the change in the mineral composition of clay seals with depth is of great interest. Examination of data on the composition of clay seals of types I and II occurring at various depths has shown that a regular decrease in the contents of swelling minerals is observed with the growth of the depth of occurrence and age of clay deposits (Fig. 6.2). In young sediments occurring at depths shallower than 100–200 m, an obvious predominance of swelling minerals (montmorillonite and mixed-layer minerals of the montmorillonite-illite type) is observed in the composition of clay seals of types I and II; their content is almost 10 times as high as that of non-swelling minerals.

As the clay deposits subside, the ratio of swelling to non-swelling minerals (K_m) decreases almost logarithmically and is slightly less than 2 at depths of 2500 m (Fig. 6.2). A correlation analysis allows one to establish the following dependence for the plot obtained:

$$K_m = 25.868 - 3.025 \times \ln h; \quad R = 0.89 \tag{6.2}$$

The obtained relation may be explained by two causes. Firstly, as has been already noted in Chapter 1, as sediments subside in the course of lithogenesis, a regular transformation of clay minerals is observed, with a trend of montmorillonite transformation at first into the mixed-layer phases and then into non-swelling

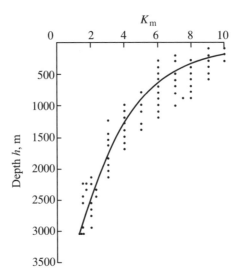

Fig. 6.2. Change in the ratio of swelling and non-swelling mineral contents (K_m = swell/unswell) with depth for the seals of classes I and II.

minerals of the illite and chlorite types. The second cause may be the evolution of source areas and sedimentation basins, when at the initial stages of sedimentation basin development (lower parts of the section), a somewhat greater amount of silt and clay particles of non-swelling minerals is supplied by turbidity currents into the deep-water zone. With further evolution of the sedimentation basin (upper parts of the section), the clay material is supplied mainly from the less altered lower horizons of weathering crusts enriched in montmorillonite. In both cases (lithogenesis and sedimentary basin evolution), a gradual enrichment of sediments in non-swelling minerals is observed down the section.

The decrease in the contents of swelling minerals promotes deterioration of the screening properties of clays with depth. However, for the seals of types I and II, the rates of this process are relatively slow: even at depths about 2.5 km and more, the clays remain obviously montmorillonite-mixed-layer in composition, which suggests the preservation of their high isolating properties.

The exchange capacity of clay deposits is closely associated with the change in their granulometric and mineral composition. An analysis of clays from various sedimentation basins shows a general tendency towards the lowering of their exchange capacity with depth (Fig. 6.3). The correlation dependence for the exchange capacity as a function of depth, in mg-equiv./100 g of dry rock specimen, has the following expression:

$$E_c = 61.076 - 5.441 \times \ln h; \quad R = 0.89 \tag{6.3}$$

The obtained data are explained by the fact that both the lowering of sediment dispersion and the decrease in the contents of swelling minerals lead to the

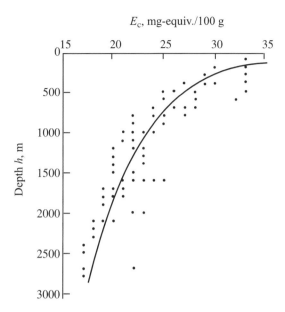

Fig. 6.3. Change in the exchange capacity (E_c) with depth for the seals of classes I and II.

reduction of the unit surface of mineral particles and to the inevitable decrease in their exchange capacity.

The change in the composition of exchange cations occurring under the influence of lithogenetic processes is of no less significance for the estimation of seal properties. The initial composition of the exchange complex of sediments is determined by the facies-geochemical conditions of their accumulation. For the sediments that form seals of types I and II, the prevalent cation is sodium, whose content amounts to 60% of the sum of exchange cations [107]. As the depth grows, a gradual decrease in the quantity of sodium and a concurrent increase in that of calcium occur. In spite of this, the content of sodium remains rather high even at considerable depths. An evidence for this is the ratio of the amount of bivalent cations (calcium, magnesium) to that of sodium, which at a depth of about 2.5 km does not exceed 0.4–0.7.

The reduction of the exchange capacity and sodium content promotes degradation of the screening properties of clay sediments. For the seals of types I and II, this tendency is not critical, as the clay rocks making up these seals preserve the high exchange capacity and prevalence of sodium ion in the exchange complex up to significant depths.

6.1.2.   FACIES OF THE MIDDLE SHELF AND PERIPHERAL
         PART OF DELTA
Clay sediments of this facies accumulate in the medium water depth environments (100–200 m) of shelf and peripheral zone of delta under the conditions of

weak bottom water hydrodynamics. In the sedimentation process in this zone, along with clay material, also silt material participates, which is connected with the distant effect of a submarine delta and the activity of turbid flows.

The sediments are characterized by the homogeneity and persistence of lithological composition, horizontal bedding with elements of cross-undulation and thin cross bedding. The bedding is accentuated by silt material and fine grained organic matter (partly of the sapropel type). The content of sand-silt material is, as a rule, 10–20% amounting to 40% in the marginal zones.

The mineral composition is represented by mixed-layer varieties of the illite-montmorillonite type with the amount of swelling layers from 20–40 to 40–60% and by illite. The prevalent size of particles is 0.1–0.2 μm. Montmorillonite, kaolinite, chlorite are contained in lesser quantities.

The absorbed complex in clay sediments of the medium-depth shelf is characterized by some decrease in the contents of sodium and magnesium cations in comparison with sediments of the deep-water zone and by the increase in the content of calcium cations. The absorption capacity averages 18–22 mg-equiv./100 g.

The clay sediments of the indicated facies form in the process of lithogenesis the seals of classes III and IV and are characterized by relatively high screening properties: they are partially permeable to gas and impermeable to oil.

As the sediments subside, there is a gradual increase in the content of sand-silt material: at a depth of 1.5–2.0 km, the ratio of sand-silt to clay particle contents is about 0.2, while at a depth of 2.5–3.0 km this ratio attains 0.3–0.35 (Fig. 6.4). The obtained relationship is described by the following correlation equation:

$$K_{sa/c} = 1.386 + 0.217 \times \ln h; \quad R = 0.82 \tag{6.4}$$

Such a regularity is connected with the greater activity of a submarine delta and associated turbid flows at the earlier stages of basin development (lower part of the section) and with the following gradual attenuation of its activity (upper part of the section). Besides, some reduction in the sediment dispersion is possible in the course of lithogenesis resulting from the processes of dissolution, recrystallization, aggregation, and cementation.

A characteristic feature of clay mineral composition is the change in the ratio of swelling to non-swelling minerals (K_m) with depth. As is seen from Figure 6.5, at a depth of 1200–1500 m, the quantity of swelling varieties is approximately two times as high as that of non-swelling minerals. At a depth of about 2500 m, the contents of swelling and non-swelling minerals become practically equal. The obtained relationship is described by the following correlation equation:

$$K_m = 8.694 - 0.959 \times \ln h; \quad R = 0.72 \tag{6.5}$$

The graph reflects a regular increase in the contents of non-swelling minerals with depth, which is explained by the transformation of swelling components into non-swelling ones and by the growth of illite and chlorite contents. This

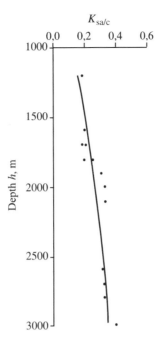

Fig. 6.4. Change in the ratio of sand-silt and clay fraction contents ($K_{sa/c}$ = (sand + silt)/clay) with depth for the seals of classes III–IV.

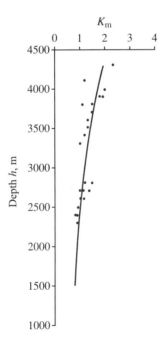

Fig. 6.5. Change in the ratio of swelling and non-swelling mineral contents (K_m = swell/unswell) with depth for the seals of classes III–IV.

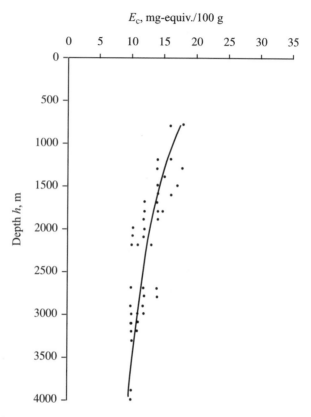

Fig. 6.6. Change in the exchange capacity (E_c) with depth for the seals of classes III–IV.

trend may be partly enhanced by the gradual weakening of turbidity current activity with sedimentary basin evolution and by the increase in the contents of fine grained swelling minerals in the upper part of the section. It is important to note that, in spite of the well pronounced tendency of a decrease in swelling components with depth, the content of mixed-layer minerals and montmorillonite in the seals of classes III and IV remains dominant up to depths of 2.5–3.0 km, after which the non-swelling minerals begin to prevail. On this basis, the seals under consideration are believed to preserve their rather high screening properties up to the depths corresponding to the middle catagenesis subzone.

The change in the mineral composition and dispersion (grain size distribution) of clay sediments is reflected in a regular decrease in their exchange capacity, whose value drops from 16–18 mg-equiv./100 g at a depth of 700–800 m to 10 mg-equiv./100 g at a depth of 2.5–3.0 km (Fig. 6.6). An equation for the curve of exchange capacity vs. sediment burial depth has the following form:

$$E_c = 50.765 - 4.998 \times \ln h; \quad R = 0.85 \tag{6.6}$$

The same as for the deep-water sediments, the change in the exchange capacity with depth for the sediments of the considered facies is accompanied by an increase in the contents of calcium and magnesium in the exchange complex at the expense of decrease in the sodium content.

6.1.3. FACIES OF THE SHALLOW-WATER SHELF AND COASTAL ZONE

The sediments of the indicated facies form in the shallow (<100 m) shelf and coastal shallow-water environments characterized by the enhanced dynamic mobility and activity.

The initial sediment is lithologically inconsistent, with the prevalence of lenticular layers of sand and silt material. Such a sediment composition is explained by the immediate influence on the sedimentation of the subaerial and submarine delta, tidal processes, and near-shore currents. According to their granulometric composition, the sediments are classified as clayey silts consisting mainly of fine (0.01–0.005 mm) and coarse (0.05–0.01 mm) silt fractions.

A considerable amount of humus-type organic matter in the form of fragments of higher plant tissue is present in the sediments, which together with the admixture of silt and sand particles accentuates the unidirectional and not unidirectional cross bedding reflecting the active hydrodynamics of aqueous medium.

Kaolinite prevails in the clay component composition, illite and chlorite are also present. Mixed-layer (illite-montmorillonite) minerals are contained in extremely restricted quantities and mainly in finer sediments accumulated farther from the shore. The prevalence of kaolinite is determined by the conditions of mechanical sedimentary differentiation, when the largest particles of kaolinite and its aggregates (1 μm and more in size) precipitate in the coastal zone, with particles of illite (average size about 0.5 μm) and chlorite (0.3–0.5 μm) settling down farther offshore.

In the exchange complex, calcium and magnesium are dominant. The exchangeable sodium is present in close correlation with calcium. The ratio of exchangeable magnesium to its sum with calcium does not exceed 0.3. The absorption capacity value is insignificant and does not exceed 10–12 mg-equiv./100 g of a specimen.

The clay sediments deposited in the shallow-water shelf and coastal environments form in the process of lithogenesis the seals of classes V–VII characterized by low screening properties: complete permeability to gas and partial or complete permeability to oil. The depositional environments of these sediments are favorable for the accumulation in them of a considerable amount of sand-silt particles, whose content gradually increases with the growth of sediment burial depth (Fig. 6.7). Thus, at a depth of 600 m, the ratio of sand-silt to clay fractions is 0.4, while at a depth of 2.0–2.5 km it is close to 1, evidencing that at these depths the contents of sand-silt and clay particles are approximately equal. The correlation equation for the graph shown in Figure 6.7 is as follows:

$$K_{sa/c} = 2.439 + 0.448 \times \ln h; \quad R = 0.96 \tag{6.7}$$

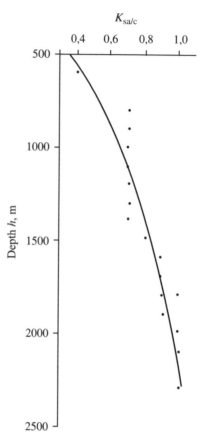

Fig. 6.7. Change in the ratio of sand-silt and clay fraction contents ($K_{sa/c}$ = (sand + silt)/clay) with depth for the seals of classes V–VII.

A considerable change in the grain size distribution of the V–VII class seals with depth is explained not only by the evolution of source areas on the continent during the sedimentary basin formation, but also by the intense development of recrystallization and cementation processes accompanied by change in the initial grain size distribution of sedimentary material.

A characteristic feature of the mineral composition of the V–VII class clay seals is the predominance of non-swelling minerals. Thus, at depths of 500–700 m the ratio of swelling to non-swelling mineral contents is about 1, for a depth of 2 km it attains 0.5 and equals 0.1 for a depth around 4 km (Fig. 6.8). The obtained dependence is described by the following correlation equation:

$$K_m = 4.949 - 0.574 \times \ln h; \quad R = 0.85 \tag{6.8}$$

It is obvious that the decrease in grain size distribution and the enrichment of sediments in non-swelling clay minerals cause quick degradation of the isolating

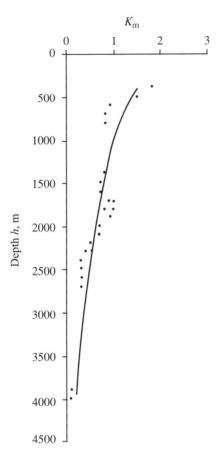

Fig. 6.8. Change in the ratio of swelling and non-swelling mineral contents (K_m = swell/unswell) with depth for the seals of classes V–VII.

properties of this type of seals with depth. This is also evidenced by the low value of the exchange capacity of sediments under consideration, which is 9–10 mg-equiv./100 g at a depth of 500 m and about 5 mg-equiv./100 g at a depth of 2.5 km (Fig. 6.9). The correlation equation for the exchange capacity as a function of depth has the following form:

$$E_c = 24.639 - 2.470 \times \ln h; \quad R = 0.78 \tag{6.9}$$

In the exchange complex composition, the cations of magnesium and calcium prevail, which enrich the sediments in the vicinity of deltas and coastal shallow-water zones. With depth, the quantity of bivalent cations in the exchange complex increases, which intensifies the processes of aggregation in clays and favors the increase in their permeability.

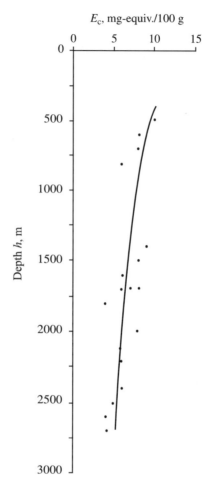

Fig. 6.9. Change in the exchange capacity (E_c) with depth for the seals of classes V–VII.

6.2.   A COMPARATIVE CHARACTERISTIC OF THE COMPOSITION OF CLAY SEALS OF VARIOUS FACIES TYPES

As has been shown above, the sediments accumulating in different zones of a sedimentation basin and forming seals with different screening properties differ in the granulometric composition, the composition of clay components and their quantitative proportions, the absorption capacity and absorbed complex composition. The indicated characteristics change with depth in a different way for sediments accumulated in different facies zones.

Thus, the ratio of sand-silt to clay material for the seals of classes I and II (Fig. 6.10, curve 1) changes insignificantly within a depth range from 20 m to 1000 m, where it grows from 0.01 to 0.1; downward to 3000 m this ratio increases to 0.2. The curve for the medium-depth shelf sediments forming the seals of classes III and IV (Fig. 6.10, curve 2) shows a more significant change

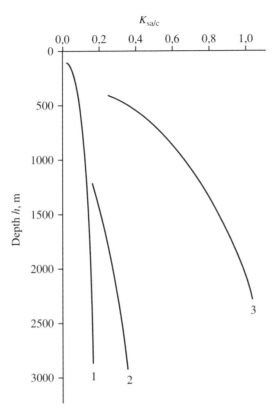

Fig. 6.10. The character of change in the ratio of sand-silt and clay fraction contents ($K_{sa/c}$ = (sand + silt)/clay) with depth for clay seals of different facies types: 1 – facies of the deep-water shelf and adjacent sea basin (seals of classes I and II); 2 – facies of the medium-depth shelf and peripheral part of submarine delta (seals of classes III–IV); 3 – facies of the shallow-water shelf and coastal zone (seals of classes V–VII).

in the sediment grain size distribution with depth. The curve has a slightly concave character. The ratio of the sand-silt fraction to clay fraction changes within a depth range of 1500–3000 m from 0.2 to 0.35. For the seals of classes V–VII (Fig. 6.10, curve 3), the change in grain size distribution with depth is still more pronounced. The ratio value $K_{sa/c}$ increases for them to 1 (at 2400 m). The change in the ratio of swelling to non-swelling mineral contents for all of the seal varieties is presented in Figure 6.11.

For all sediment types, a decrease with depth in the ratio of swelling minerals to non-swelling varieties (K_m) is registered. For the seals of classes I and II (Fig. 6.11, curve 1), a significant prevalence of swelling minerals is characteristic. In the upper part of the section (depth 200 m), the value of K_m for them is 10. As the depth increases, K_m quickly diminishes, but in spite of this, even at depths of 2.0–2.5 km the swelling minerals noticeably prevail over the non-swelling varieties ($K_m \cong 2$).

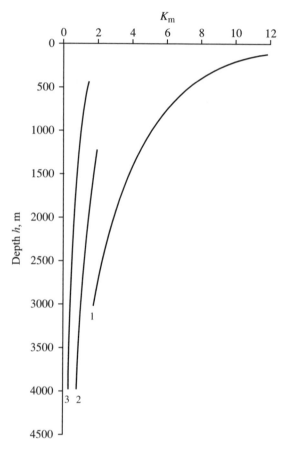

Fig. 6.11. The character of change in the ratio of swelling and non-swelling mineral contents (K_m = swell/unswell) with depth for clay seals of different facies types (for legend see Fig. 6.10).

For the seals of classes III and IV, the prevalence of swelling over non-swelling mineral contents ($K_m \approx 2$) is observed in the upper parts of the section (Fig. 6.11, curve 2). At a depth around 2.5 km, the contents of both groups of minerals become equal, and at greater depths the non-swelling mineral varieties become dominant.

The seals of classes V–VII, each beginning with a depth of 1000 m, are characterized by the prevalence of non-swelling minerals, whose content grows quickly with depth (Fig. 6.11, curve 3).

The values of the exchange capacity of sediments are closely connected with the character of changes in their grain size distribution and in the quantity of swelling minerals with depth. As is seen from Figure 6.12, the value of E_c for all the three facies varieties of clay seals decreases regularly with depth. The differences are only in the absolute values of the change of this parameter.

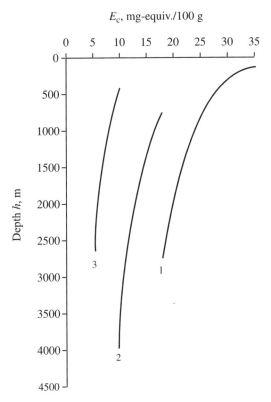

Fig. 6.12. The character of change in the exchange capacity (E_c) with depth for clay seals of different facies types (for legend see Fig. 6.10).

Thus, for the seals of classes I and II (curve 1), the value of E_c in the depth interval from 200–300 m to 2.5–3.0 km diminishes from 35 to 18 mg-equiv./100 g, for the seals of classes III and IV (curve 2) the value of E_c changes from 17 mg-equiv./100 g (at a depth of 800 m) to 10 mg-equiv./100 g (4000 m). For the seals of classes V–VII (curve 3), the exchange capacity decreases in the depth interval of 400–2700 m from 10 to 5 mg-equiv./100 g.

7

Structure and Properties of the Facies Types of Clay Seals

7.1. THE INFLUENCE OF DEPOSITIONAL ENVIRONMENTS ON THE STRUCTURE FORMATION

General problems of structure formation in clay rocks have been stated in Chapter 1. Considered below will be the processes of clay sediment deposition and primary microstructure formation in relation to certain depositional environments, as well as the further transformation of the formed sediments in the course of lithogenesis.

7.1.1. FACIES OF THE DEEP-WATER SHELF AND ADJACENT SEA BASIN

As has been already noted, the formation of deep-water clay sediments occurs in the deep-water part of sea shelf far from the shoreline under the conditions of a relatively quiet state of the bottom water layer and restricted supply of clastic sand and silt material. The sediments deposited here form in the course of lithogenesis the seals of types I and II, impermeable to oils and hydrocarbon gases. The mineral composition of the sediments is dominated by montmorillonite and mixed-layer minerals of the montmorillonite-illite type with swelling layers making up from 40–60% to 60–80%. Both of the minerals are represented by fine grained particles 0.1–0.2 μm in size. Illite, kaolinite, chlorite may be present as admixtures. The contents of sand-silt inclusions does not exceed 10%. In the absorbed complex, Na^+ dominates making up no less than 60% of the sum of exchange cations.

Montmorillonite and mixed-layer minerals, which are widely spread among the sediments of deep-water facies, exert an extremely great influence on the structure formation in these sediments. The high wetting ability of external and internal surfaces of these minerals causes their ability to retain considerable quantities of adsorptionally bound and osmotic water. The bound water forms on the surface of minerals hydrate films, whose thickness may exceed that of the particles [8, 9, 51, 217–221].

The hydrate films of bound water, having specific structural–mechanical properties, play the role of a stabilizing factor not allowing the particles in a

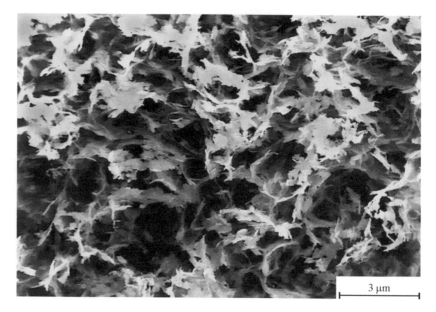

3 μm

Fig. 7.1. Microstructure of a Na-montmorillonite sediment.

water medium to approach each other at the distance of distant or close potential minimum operation and to form large aggregates [6]. Therefore, the settling of montmorillonite and mixed-layer mineral particles proceeds very slowly under the quiet conditions of the deep-water part of a sedimentary basin. In this process, the particles settle either separately or in the form of ultramicroaggregates, and only in the near-bottom part of the basin, with an increase in their concentration, they lose stability and form fine microaggregates of a leaf-like shape that come into interaction with each other (coagulate) due to molecular forces to form a uniform fine-porous (small-honeycomb) sediment . The interaction of microaggregates occurs mainly according to the face-to-face or face-to-edge type with low angles of particle inclination, which creates a weak axial orientation in the sediment. Figure 7.1 shows such a sediment obtained by us in precipitating Na-montmorillonite.

The formed sediment has porosity (80–85%) caused by the presence of fine micropores (0.1–1 μm) and ultramicropores (<0.1 μm). The high wetting ability of minerals and the small pore sizes are the reasons why a considerable part of water (up to 40–60%) contained in the sediment pores proves to be in the bound state and is tightly retained on the external surfaces and in the internal (swelling) interlayer space of montmorillonite and mixed-layer minerals.

The described mechanism of sediment formation is much promoted by the dominance of Na^+ cation in the exchange complex and by the presence of high-molecular organic matter in the sediment. The presence of exchangeable Na^+ causes an increase of the diffuse ionic layer and intensifies the stabilization of clay particles. Humified organic matter is being actively adsorbed on the surface of montmorillonite, this also resulting in the increase of the wetting ability and stability of clay particles in a water medium.

An increase of salt concentration in the aqueous dispersion and the appearance of multivalent cations in the exchange complex of montmorillonite reduce the stability of its particles and lead to the enlargement of microaggregates. In this process, the sediment microstructure acquires at first the small-honeycomb form and then the medium- and even large-honeycomb form.

Homogeneous clay sediments of montmorillonite and mixed-layer minerals have a great reserve of surface energy, and so the physicochemical and biochemical processes tending to lower the free surface energy and to cause spontaneous compaction develop in them already at the very beginning of lithogenesis (at the early diagenesis substage).

As the sediment overburden grows, the gravitational pressure becomes the main factor of compaction. Free pore water is squeezed out in the course of diagenesis and, already by the end of the late diagenesis substage, further compaction of the sediment begins to be controlled by the conditions of bound water removal. Due to the specific structural–mechanical properties and enhanced shearing strength of bound water, its squeezing out of a porous sediment may only proceed under an overpressure. So the process of dewatering of a sediment composed of swelling minerals becomes, in its physical nature, mainly of the diffusion rather than filtration type at the end of the late diagenesis substage and at the early catagenesis substage. The diffusional moisture transfer develops extremely slowly, so that coagulation contacts are preserved in the sediment throughout the early and the larger part of the middle catagenesis substages. The presence of coagulation contacts at such great depths (up to 3000 m and more) is accounted for by the low effective stress at the contacts owing to the high dispersion of the system allowing the preservation of thin hydrate films in the contact clearance. In addition, the extremely low supply of new elements due to the absence of filtration transfer of substance by pore water impedes the development of cementation in them. This is also the reason why the processes of ion exchange and mineral transformations are retarded in deep-water sediments. As has been shown in the previous chapter, a characteristic feature of these sediments is a small change with depth in the ratio of swelling to non-swelling minerals (M–I + M/K + Chl), as well as in the ratio of exchangeable Na^+/Ca^{2+}.

At the end of the middle and beginning of the late catagenesis substages, the mechanism of moisture transfer in the deep-water sedimentary sequence changes. With the transition of bound water into free state at temperatures of more than 65–70 °C, the conditions for the filtration squeezing-out of pore moisture are created. By this time, however, the sediment pores become so fine that no sharp increase in the compressibility of the sequence results from this. Nevertheless, the transition of adsorptionally bound water into free state promotes the further rock compaction at depths as great as 5–10 km. It is possible that sediment compaction at these depths becomes rheological in its nature and is connected with plastic deformations of particles at contacts. An important matter is that, thanks to the high homogeneity of the rock structure, no local stress concentration leading to microfracturing arises in it. Therefore, the clay rocks of montmorillonite and mixed-layer mineral composition are capable of preserving their high isolating properties at such great depths.

The structure of a deep-water sedimentary sequence, along with the mineral composition and microstructure, exerts a great influence on the slow progressive compaction of clays having a montmorillonite-mixed-layer composition. As already noted, the depositional environments of deep-water sediments favor the formation of rather thick homogeneous deposits devoid of interlayers of coarser material. Such a structure of the sequence makes the process of pore water squeezing-out extremely difficult and slow. Because of the absence of drainage interlayers and the longer paths of pore water migration, the process of dehydration and, hence, compaction of homogeneous fine grained sediments, develops extremely slowly, which also contributes to the preservation of their isolating properties at great depths.

7.1.2.   FACIES OF THE MIDDLE SHELF AND PERIPHERAL PART OF DELTA

Clay sediments deposited in the middle-shelf and peripheral part of delta environments form in lithogenesis the seals of types III and IV impermeable to oils and partly permeable to gases. The mineral composition of the sediments is dominated by illite and mixed-layer minerals (of the illite-montmorillonite type) with the ratio of swelling to non-swelling layers, as a rule, less than 40%. In the upper parts of the section, both minerals are contained in approximately equal proportions, while in deeper parts the predominance of illite becomes apparent. Montmorillonite together with kaolinite and chlorite are present as admixtures. The clay sediments contain sand-silt grains making up from 20% to 30–40%. In the absorbed complex, along with Na^+, also Ca^{2+} is widespread.

A noticeable influence on the structure formation in clay sediments of this facies is exerted by illite and little swelling mixed-layer varieties prevailing in them. Both minerals have considerably lower wetting ability and stability in water suspensions as compared to montmorillonite. Therefore, when present in suspension in a water medium, they are capable of interacting to form relatively dense microaggregates of anisometric shape from few to 10 μm and more in size. The formed microaggregates precipitate, which prevents them from being transported to the deep-water part of the basin and causes their accumulation in the sediments formed in the middle-shelf and peripheral part of delta environments.

When settled on the basin bottom, the microaggregates come to interact with each other to form the open medium- and large-honeycomb microstructure shown on model II-f (Fig. 1.10). The cell size in such a structure may attain 10–12 μm in diameter, the configuration of pores is mainly isometric, rounded. The sediment with honeycomb microstructure has high porosity amounting to 85–90%. The pore space is heterogeneous: along with large micropores-cells, there are small (1–10 μm) intermicroaggregate and fine (0.1–1 μm) intramicroaggregate micropores in the sediment. The latter are commonly anisometric (elongated or wedge-shaped) in shape. The water content of sediments is much higher than their water content at the liquid limit. The bulk of the pore moisture

Fig. 7.2. Microstructure of a marine sediment composed of illite.

is in free state and contained in micropores-cells, the part of bound water makes up no more than 20–25% of the total water content.

An illite sediment having all the above-described characteristic features has been first obtained artificially by N.R. O'Brien [70]. For natural sediments of illite compositions, the honeycomb microstructure has been described by R. Push [71], R. Push and M. Arnold [222], V.I. Osipov and V.N. Sokolov [68] (Fig. 7.2).

The honeycomb microstructure is the optimal composition for creating the volume structure and lowering the surface energy of a dispersed system at the minimal solid phase concentration. The formation of such a microstructure is connected with specific interaction of microaggregates with each other according to the face-to-edge type. Development of such interactions is accounted for by the fact that the edges of individual particles and microaggregates are less stabilized by hydrate films, which leads to their destabilization and coagulation [6, 171, 223, 224]. When two aggregates draw together with their not stabilized parts, a contact of the edge-to-edge type forms, and when the unstable part of one microaggregate draws near the stabilized surface of another microaggregate, this results in the formation of the face-to-edge type of coagulation contact.

A growth of salt concentration in the pore solution and the prevalence of multivalent cations in the exchange complex increase the number of not stabilized portions of clay mineral surface promoting the enlargement of their microaggregates. A sediment formed under such conditions has extremely high porosity and a characteristic large-honeycomb microstructure [51, 69, 225].

The compaction of clay sediments of the illite-mixed-layer composition in the course of lithogenesis has its characteristic features making it different from the compaction of sediments composed of fine grained swelling minerals. These features are caused by the existence of large-honeycomb open microstructure in such sediments and by the lesser content of bound water in them.

Under the influence of at first physicochemical processes and then the increasing gravitational pressure, the heterogeneous large-honeycomb micro-structure of sediments begins to compact intensely already at the early diagenesis substage. The predominance of distant coagulation contacts between microaggregates causes the high mobility of the microstructure: the compaction occurs at the expense of mutual displacement of microaggregates and destruction of large intermicroaggregate pores-cells with their disintegration into smaller micropores. The presence in the sediment of up to 30–40% of sand-silt material does not influence much the compaction, as the clastic material is submerged into clay matrix and has no mutual contacts (does not form a three-dimensional structural framework). The compaction of microstructure is accompanied by the squeezing-out of free pore water from large micropores and its migration to the drainage horizons of the sequence.

The free gravitational compaction of the sediment continues till the middle of the early catagenesis subzone (depth 1200–1300 m). By this moment, the rock porosity becomes equal to 30–35%, which is noticeably less in comparison with the porosity of sediments composed of fine grained swelling minerals. The pore space structure undergoes significant alterations: large micropores and a considerable part of small (1–10 μm) intermicroaggregate micropores practically disappear, the content of fine (0.1–1 μm) micropores increases.

The accelerated compaction of sediments at this substage of lithogenesis is explained by the presence in them of large open pores mainly filled with easily squeezed out free water and by the low strength of the structure due to the lesser number of contacts in the sediment volume unit as compared to the structure composed of fine grained minerals and their ultramicroaggregates. Besides, the compaction is noticeably influenced by the lesser homogeneity of the sediments of the middle shelf and peripheral part of delta in comparison with the deep-water sediments. This creates more favorable conditions for the removal and drainage of water squeezed out of the sediment.

At the end of diagenesis – beginning of the early catagenesis substage, the squeezing-out of free water is completed and further dehydration of the sediment becomes possible only at the expense of osmotic water removal. Due to the enhanced viscosity of osmotic water and development of the counteracting process of diffusional mass transfer, the process of compaction becomes somewhat less intense. Diffusion of weakly hydrated cations (K^+, Si^{4+} and others) towards the clay sequence begins, which creates conditions for the beginning of hydromicatization of mixed-layer minerals.

As the osmotic water gets squeezed out at the beginning of the middle catagenesis substage, the process of cementation starts, which leads to the formation of at first transitional and then phase contacts. In spite of the increase in the strength of structural bonds, the rock continues to be compacted at the expanse of rheological processes without breaking its continuity.

The character of compaction and deformation of rocks of hydromica and mixed-layer mineral composition changes noticeably when the process of adsorptionally bound to free water transformation begins at the end of the middle catagenesis substage. The bound water transformation is accompanied by the lessening of Rebinder's effect – the adsorptional reduction of rock strength, which results in the noticeable increase in the strength and brittleness of rocks. Therefore, the further deformation of rocks during compaction begins to have the character of brittle destruction and causes the appearance of microfractures. The development of microfracturing intensifies at the end of the middle and at the late catagenesis substages with the beginning of rock argillization which leads to the formation of platy pseudocrystals and increases the heterogeneity of structure and structural bonds.

7.1.3. FACIES OF THE SHALLOW-WATER SHELF AND COASTAL ZONE

The sediments of the type under consideration form in the shallow shelf and coastal agitated shallow-water environments under the conditions of enhanced dynamic activity of near-shore currents. They make up the seals of classes V, VI, and VII permeable to gases and partially impermeable to oils. The principal mineral components of the sediments are the non-swelling minerals kaolinite, illite, chlorite. Among the swelling varieties, the presence of mixed-layer minerals of the montmorillonite-illite type is possible, but their content quickly decreases with depth. The sediments contain a considerable amount of sand and silt grains whose percentage varies from 30–40% to 70–80%. Calcium and magnesium are dominant in the exchange complex. The sediment structure is characterized by strong heterogeneity: clay sediments alternate with interbeds and lenses of sand and silt material.

The structure formation in the shallow-water clay sediments is determined by several basic factors, and namely: (1) the low wetting ability of non-swelling minerals and their tendency in coagulation to form large-honeycomb sediments with rather low structural strength, easily compacted even under small loading; (2) the high contents of clastic grains having sand-silt sizes, capable at a certain compaction stage of creating their own structural framework and impeding the compaction; (3) at last, the high heterogeneity of the sedimentary sequence, the presence in it of interlayers and lenses of coarser material, which create favorable conditions for the drainage of pore waters contained in clay sediments and for active geochemical exchange between the sedimentary sequence and the underground waters.

Clay minerals of the kaolinite and chlorite group are even less stable in natural dispersions than illite and weakly swelling mixed-layer varieties. This is explained by the relatively large size of their particles (0.5–10 μm) and low wetting ability. Because of the poorly developed diffuse layer and small thickness of surface hydrate films, the particles of these minerals are capable of aggregating with the formation of close coagulation contacts and thus of further increasing their sizes. In the formed microaggregates, the particles interact according to the basis–basis type. The most specific in this respect is kaolinite which forms flat microaggregates of the «shifted pack of cards» type [69, 70].

The large particle sizes of kaolinite, chlorite, and their aggregates promote the settling of these minerals at an early stage of sedimentogenesis. So, their accumulation proceeds in the shallow shelf and coastal shallow-water environments, where they precipitate together with the sand-silt material and with the largest microaggregates of illite.

In the near-bottom part of the basin, the microaggregates of kaolinite and chlorite begin interacting with each other to form a large-honeycomb microstructure similar to the one shown on scheme II-e (Fig. 1.10). The interaction of microaggregates in such a microstructure is of the basis–basis type or of the face-to-edge type with small angles of microaggregate inclination towards each other. Such a microstructure for kaolinite sediments has been for the first time obtained under laboratory conditions by O'Brien [70], later it has been confirmed by V.I. Osipov [51] (Fig. 7.3).

The microstructure of a purely kaolinite sediment has an extremely high porosity (up to 95%). Individual pore sizes are as large as 10–18 μm. The weight water content of the sediment is 130–300%. The bulk of the pore moisture is free water, the percentage of bound water is insignificant and does not exceed 5–10%.

As has been already noted, deposition of clay minerals in the area of middle and particularly shallow shelf and coastal shallow water occurs together with clastic material of sand and silt fractions. This causes the development of heterocoagulation processes, when clay particles interact not only with each other, but also with larger clastic grains. The surface (molecular) forces of sand-silt

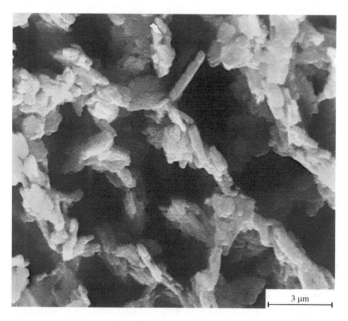

Fig. 7.3. Microstructure of a kaolinite sediment.

particles are weakly screened by hydrate films, and so the clastic grains represent some kind of adsorption centers for the more dispersed clay material [6, 171]. This causes the development of heterocoagulation in a polydispersion system, i.e. sticking of smaller clay particles to the surface of clastic grains. As a result, an aggregate of a clastic grain surrounded by stuck clay particles forms, which settles together with clay microaggregates and participates in the formation of the sediment microstructure.

Owing to the presence of dense aggregates, the initial porosity of the formed sediment is somewhat lower in comparison with the pure kaolinite sediment. The sand-silt particles, when their content in the sediment is not high (less than 40%), are submerged into the clay matrix without contacting with each other and so do not have a significant influence on the sediment compaction, for their mutual displacement and sliding occur along the clay envelops surrounding them.

The picture changes considerably with an increase in the contents of silt and sand particles in the sediment to 40% and more. This leads to further reduction of porosity in the initial sediment, which may be no more than 60–70%. At early stages of lithogenesis, such a sediment may become quickly compacted at the expense of clay matter located in the clearance between grains, which will play the role of a «lubricant». Therefore, there is a noticeable increase in sediment density already in the diagenesis zone. With further compaction, the number of immediate contacts between clastic grains gradually increases, they form their own structural framework, which begins to take effective stresses and thus to retard the sediment compaction.

So, the compaction of clay sediments formed in the shallow-water environment depends not only on the microstructure of clay matrix, but also on the contents of clastic grains. The compaction of clay matrix in such sediments develops very quickly. This is explained by the presence of large micropores practically entirely filled with free water and by the low structural viscosity of the sediment proper. Therefore, already by the end of diagenesis, the sediment in which the content of clastic grains does not exceed 40% is capable of being compacted to the porosity value of no more than 30–35%. The structural heterogeneity of shallow-water sediments, the presence in their section of interbeds and lenses of coarser material, which drains well the pore moisture, exert a great influence upon their accelerated compaction.

With an increase in the contents of sand-silt grains, the compaction of shallow-water clays slows down: after a certain degree of compaction is reached, the effective stresses are transmitted to the contacts of clastic grains and the compaction becomes dependent on the mobility of the sand-silt structural framework. The clay matter inside the framework of clastic grains remains undercompacted and preserves a considerable amount of medium and fine micropores.

Thanks to the latter peculiarity, the shallow-water clays take active part in the processes of mass exchange with solutions circulating in reservoir beds, which provides the supply of new compounds to them and the formation of cement. The cementation of rock starts to develop already at the beginning of the early catagenesis substage, and its further compaction proceeds not only under the action of gravitational pressure, but also at the expense of new compounds educed in the pore space. The early cementation of clays results in the increase of rock strength and imparts high rigidity (brittleness) to the framework. This circumstance, as well as the presence of structural hetero-geneities, favors the appearance of stress concentrators and the brittle local breaking of microstructure with the formation of microfractures. Thus, already at the beginning of the middle catagenesis substage, the shallow-water clays having porosity of 10–15% may lose their screening properties.

7.2.   POROSITY

As has been noted in Chapter 3, the compaction of clay sediments in the course of lithogenesis and the formation of their screening properties is a complex physicochemical and mechanical process depending on the composition, homo-geneity, and thickness of the sedimentary sequence, the geodynamic, pressure–temperature, and geochemical conditions of compaction. Various combinations of these factors result in a variety of possible courses of sediment compaction, and hence – in wide scattering of density and porosity values with depth [226].

The most important factors in the compaction of sediments are their initial composition and the structure of the forming sedimentary sequence, which are determined by the facies environments of sediment accumulation. The mineral composition of sediments, the composition of pore waters and exchange complex, the composition and content of organic matter, as well as the homogeneity and thickness of sediment, the presence in it of interbeds and lenses of coarser mate-rial depend on these environments. The initial composition influences the mineral transformations and development of physicochemical processes in the sediment in the course of its lithogenesis. Therefore, in sedimentary oil- and gas-bearing basins comprising various facies complexes of clay rocks (deep-water, relatively deep-water, and shallow-water), different trends of porosity change with depth will be observed. Indeed, the earlier discussed data (Figs. 3.3–3.5) obtained for different sedimentary basins evidence that every basin is characterized by its own regularity in the change of clay sediment porosity along the section.

This circumstance has been entirely disregarded by such well-known researchers as J. Weller [130] and N.B. Vassoevich [148] who have proposed the averaged clay density vs. depth relationships (Fig. 7.4). The data obtained by them do not allow prediction of the screening properties of seals, as they do not take into account the specific conditions of compaction and screening property formation of clay sediments belonging to different facies type.

It follows from the aforesaid that, when studying the dependence of poros-ity on the depth of clay sediment burial, the clays of similar composition should be considered, i.e. those formed in the same depositional environments. Such an approach permits one to reduce the scattering of porosity values and to reveal the regularities in the changes occurring in each specific facies com-plex of clay sediments. Then, based on the facies analysis of the basin and on the regularities obtained, it is possible to predict the change in clay porosity within different parts of the section [227, 228].

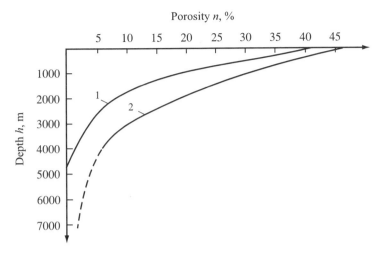

Fig. 7.4. Generalized curves showing the change in total porosity of clay sediments with depth: 1 – after J. Weller [130]; 2 – after N.B. Vassoevich [148].

7.2.1.   COMPACTION OF THE FACIES TYPES OF CLAY SEALS

Proceeding from the available published data, let us consider the change of porosity with depth for various facies types of clay sediments.

The change of porosity with depth for a deep-water facies complex may be considered with reference to the sections of the West Siberian and Peri-Caspian sedimentary basins. In the West-Siberian basin, deep-water clays occur at five depth levels (see Fig. 5.1). These are the 800–1000 m, 1100–1200 m, 1500–1600 m, and 2000–2300 m intervals. A porosity-depth dependence plotted for these intervals shows that at a depth of 600 m porosity is 30% and at a depth of 1000 m it is 23% (Fig. 7.5). The prevalent mineral in the clay composition in this depth interval is montmorillonite. With the further increase in depth, porosity decreases to 21% (1200 m), 17% (1600 m), 12% (2100 m), 7% (2800 m). At depths of 1500–1600 m, mixed-layer (montmorillonite–hydromica) minerals with 40–80% of swelling layers appear in the clays; at depths of 2000–2100 m they become dominant. Concurrently, the quantity of swelling interlayers in the structure of mixed-layer minerals decreases. It is possible that the thermodynamic process of conversion of adsorptionally bound water to free water, which develops at these depths, activates the mineral transformations tending to reduce the quantity of swelling layers in the clay mineral composition.

In the Peri-Caspian basin, deep-water clays occur at studied depths up to 6000 m. A specific feature of sediments in this basin is wide development of montmorillonite in their composition. Here, the same as in the West-Siberian basin, a progressive change of porosity with depth is observed to the depth of 6000 m. Within the depth interval of 3000 to 6000 m, porosity decreases from 8 to 2%.

In other basins (Bureya, Volga-Ural, Yamal, and others), deep-water clays occur largely in the upper parts of the section at depths up to 1000–1500 m

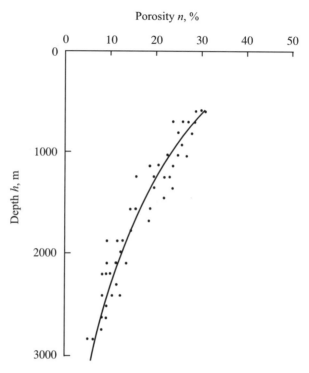

Fig. 7.5. Change in porosity with depth for clay sediments of the deep-water facies, West-Siberian basin.

(see Figs. 5.1 and 5.2). Their porosity attains 20–25%; the prevalent minerals in the clay composition are montmorillonite and mixed-layer varieties with the contents of swelling layers within the 40–80% range.

Figure 7.6 shows a generalized porosity–depth curve for clay seals of types I and II, obtained by us from an analysis of data on 20 sedimentary basins within the former Soviet Union territory. The curve is described by an equation representing a cubic polynomial as follows:

$$n = 43.584 - 0.019 \times h + 2.933 \times 10^{-6} \times h^2 - 1.508 \times 10^{-10} \times h^3$$

$$R = 0.93 \tag{7.1}$$

As can be seen from the obtained graph, the sediment becomes compacted rather quickly in the upper part of the section to depths of 500–600 m (beginning of the early catagenesis substage) at the expense of the squeezing-out of free water and closure of small intermicroaggregate pores. The sediment porosity at the end of this depth interval attains 35–40%. Then the process of compaction acquires a more smooth exponential character, which is caused by the impeded removal of bound water from the thin-micropore microstructure. Such a character of compaction is observed approximately to depths of

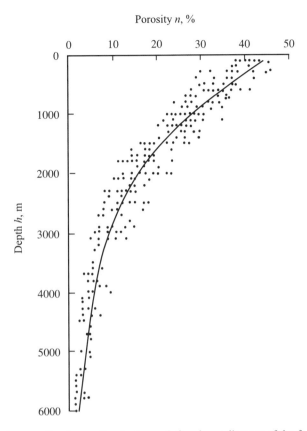

Fig. 7.6. A generalized porosity–depth graph for clay sediments of the facies of deep-water shelf and adjacent part of sea.

3000–3500 m (the end of the middle catagenesis substage), where rock porosity diminishes to 6–7%. With the transformation of adsorptionally bound water into free water, the compaction of the sequence slows down in connection with the increase in the structural bond strength and development of the rheological transformation mechanism with high viscosity of the structure. Such a deformation mechanism obviously remains dominant throughout the depth interval from 3000–3500 m to 6000 m and possibly at greater depths. The rock porosity within this interval of depths changes from 6–7% to 2%.

An important feature of the compaction process of deep-water clay seals is the predominance in this process of free geostatic compaction which develops to depths of 3000–3500 m and is determined by the mineral composition and microstructure of these rocks. Only at greater depths the process passes to the stage of impeded geostatic compaction (see Table 3.2). The second important feature is that the compaction has a character of plastic deformation within the whole depth interval under consideration and is not accompanied by the break of structural continuity, i.e. by the formation of microfractures. During free

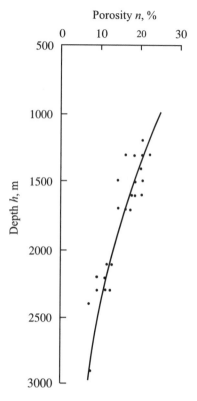

Fig. 7.7. Change in porosity with depth for clay sediments of the facies of middle shelf and peripheral part of delta, West-Siberian basin.

geostatic compaction, plastic deformations occur along the hydrate films of bound water at the contacts of particles and their microaggregates, while during the impeded gravitational compaction they occur at the expense of the plasticizing of mineral contact zones and development of the creep of structure without its destruction.

For clay sediments of the medium-depth shelf and peripheral part of delta facies, the compaction and change of porosity with depth has its peculiarities. The regularities of their compaction under natural conditions may be considered with an example of the West-Siberian sedimentary basin, where they are encountered at depths of 1200–1300 m, 1600–1700 m, 2100–2300 m, and 2800–2900 m. The clay porosity at these depths has the following values: 19, 16, 12, and 7% (Fig. 7.7).

In other considered basins (Zeya-Bureya, Volga-Ural, South-Yamal, and others), the indicated facies occur within a depth range from 1500 m to 4000 m. Everywhere, the dominant mineral in their composition is illite.

The processing of all available data on the clay sediments of middle shelf and peripheral part of delta has made it possible to obtain for them a general

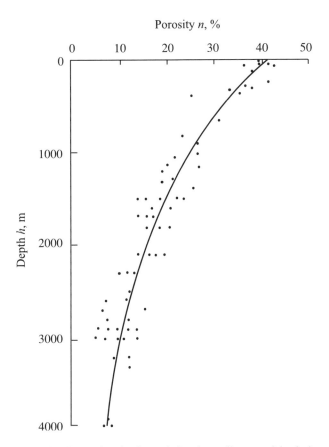

Fig. 7.8. A generalized porosity–depth graph for clay sediments of the facies of middle shelf and peripheral part of delta.

porosity-depth dependence (Fig. 7.8), which is described by the following correlation equation:

$$n = 41.127 - 0.020 \times h + 4.413 \times 10^{-6} \times h^2 - 3.571 \times 10^{-10} \times h^3$$
(7.2)

$$R = 0.91$$

The obtained data confirm the results of studies on the compaction of clays of the illite and mixed-layer composition stated above in Section 7.1.2. They show that, up to the burial depths of 1500–2000 m, these relatively deep-water sediments undergo quick compaction exceeding in its rates the compaction of clays of the montmorillonite composition. This is explained by the extremely high porosity of the sediments having the illite and mixed-layer composition and by the presence in their structure of large pores-cells filled with free pore water, which are easily destroyed under the action of even a small pressure. The lesser homogeneity of these sediments in comparison with the deep-water clays creates more favorable conditions for the removal of water in the process of its squeezing out, which also promotes their quick compaction in the diagenesis zone and in the early catagenesis subzone.

At the end of the early and beginning of the middle catagenesis substages, with porosity values being 12–20%, the character of compaction of relatively deep-water clay sediments changes and becomes less intense. One of the reasons for this is the beginning of rock cementation and increase in their structural coherence due to the rise of at first transitional and then phase contacts. The rock deformation begins to acquire a rheological character and develop at the expense of the framework creep. The whole process passes from the stage of free geostatic compaction to the stage of impeded geostatic compaction, the basis of which is, as before, the relative displacement of structural elements without the break of structural continuity and formation of microfractures.

New qualitative changes in the compaction mechanism of the sequence occur at depths of 2500–2800 m at the end of the middle catagenesis substage, with rock porosity being 8–9%. In further compaction, the recrystallization phenomena begin to play the leading part. The transition of adsorptionally bound water into free water at these depths results in a sharp increase in the brittleness of rock and in the beginning of its progressive destruction (formation of microfractures) at the places of stress concentration. The removal of tightly bound water is also assisted by the processes of mixed-layer mineral transformation into illite, which intensify with depth. Concurrently with the growth of illitization and chloritization, the processes of recrystallization and argillization begin in the rock, leading to the formation in it of large clay polycrystals and schistosity. The latter causes an increase of structural bond heterogeneity and intensification of microfracture formation along the cleavage planes.

For clay sediments of the shallow-water shelf and coastal facies, the compaction and the change of porosity also have their own peculiarities in comparison with the above described facies.

As no single section with multiple recurrence of the conditions of coastal shallow-water sedimentation is available among the basins studied by us, Figure 7.9 gives a summary porosity-depth curve for these sediments, which has been derived from the generalization of data on several sedimentary basins. The obtained dependence of porosity on the depth of rock occurrence is also described by a cubic polynomial:

$$n = 39.564 - 0.028 \times h + 9.294 \times 10^{-6} \times h^2 - 1.055 \times 10^{-10} \times h^3$$

$$\text{(7.3)}$$

$$R = 0.93$$

As can be seen from the above data, at early substages of lithogenesis, the shallow-water sediments undergo intense gravitational compaction, exceeding significantly in its intensity the compaction of the middle-shelf deposits, and especially of the deep-water sediments. Thus, by the end of diagenesis (depth $\approx$ 90 m), the sediment porosity becomes equal to 35–40%, and by the completion of the early catagenesis stage (depth $\approx$ 1000 m) it attains 10–20%. With the

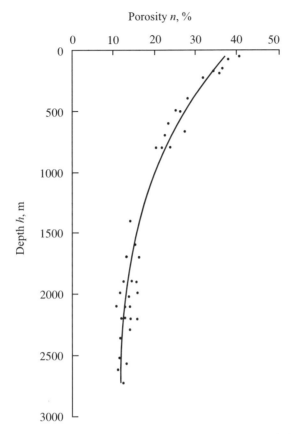

Fig. 7.9. A generalized porosity-depth graph for clay sediments of the facies of shallow shelf and coastal agitated shallow-water zone.

passage to the subzone of middle catagenesis, the rate of shallow-water sediment compaction sharply decreases, and their further compaction proceeds much more slowly than that of more clayey sediments from the middle and deep-water shelf areas.

The revealed regularity in the compaction of shallow-water clay sediments with depth may be explained by the above-indicated peculiarities of these sediments. The intense compaction at early substages of lithogenesis is connected with the low structural strength of the sediment (due to relatively large sizes of its constituent structural elements – microaggregates of kaolinite, chlorite, and sand-silt grains, allowing the formation of a restricted number of contacts in a unit volume), its high drainage ability, and the predominance in pores of unbound water easily squeezed out of the sediment. All these promote the free gravitational compaction which continues to a depth of 1200–1500 m.

When the indicated depth is reached, the character of shallow-water clay compaction undergoes significant changes. One of the reasons for it may be the formation of the structural framework of sand-silt grains. The immediate

contiguity of the grains and the increase in the effective stresses at their contacts favor the development of a conformal-regeneration structure. Another
reason is the beginning of cementation and increase in the strength of structural
bonds due to the formation of transitional and phase contacts. The process of
cementation of the sediments under discussion begins earlier than in the more
homogeneous sediments of the middle and deep-water shelf and is caused by
the relatively free circulation of pore solution and by the supply to the clay
sequence of new compounds from the contacting with it underground waters.
It should be added to the aforesaid that at these depths, changes in the mineral
composition of the sediments occur: the mixed-layer minerals disappear completely and epigenetic alterations of kaolinite and chlorite begin. As a result of
development of the indicated processes, the compaction proceeds extremely
slowly: within the depth range from 1500 m to 3000 m a decrease in porosity is
only 2–5%.

It is important that, as the cementation of the heterogeneous microstructure
of shallow-water clay sediments begins, the concentrators of stresses arise in it
(at the contacts of grains and microaggregates and around the largest micropores), which cause the break of microstructure continuity and formation of
microfractures. So, the development of compaction at the expense of framework
creep is not characteristic of these rocks, which leads to their early destruction
and loss of screening properties at depths not exceeding 1800–2000 m.

7.2.2. A COMPARATIVE ESTIMATION OF POROSITY OF
VARIOUS FACIES TYPES OF CLAY SEALS

The above-conducted analysis clearly illustrates the specificity of behavior
shown by the clay rocks formed in different depositional environments. A comparison of the compaction graphs for the three groups of clay sediments – of
deep-water, relatively deep-water, and shallow-water origin – evidences for the
presence of significant differences in the compaction of these clays in the course
of lithogenesis.

As is seen from Figure 7.10, the most intense compaction at early lithogenesis stages (the early and late diagenesis substages and the early catagenesis
substage) is observed for the shallow-water clay sediments. The relatively deep-
water deposits are compacted at these stages of lithogenesis at lower rates than
the shallow-water sediments. The deep-water clays undergo the least compaction at the beginning of lithogenesis. Thus, at a depth of ≈300 m, the porosity of shallow-water, relatively deep-water, and deep-water sediments is 32, 35,
and 40% respectively. By the end of the early – beginning of the middle catagenesis substage, the compaction of shallow-water and relatively deep-water clays
slows down, while clays of the deep-water facies continue to be progressively
compacted. At depths of approximately 2500–3000 m, the porosity of all of the
groups becomes similar and equal to 10–12%. Then the compaction of relatively deep-water and shallow-water clays begins gradually to fall behind that
of deep-water clays, and the mechanism of their compaction changes significantly: while geostatic loading continues to play an important part in the

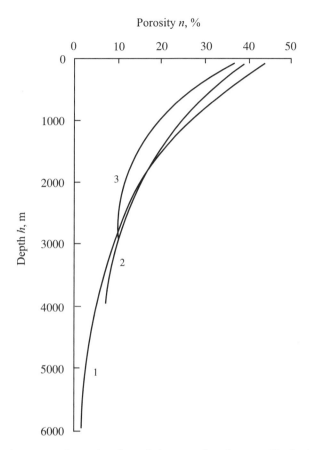

Fig. 7.10. A comparative estimation of the porosity change with depth for clay sediments formed in different depositional environments: 1 – environments of the deep-water shelf and adjacent parts of sea, 2 – environments of the middle shelf and peripheral part of delta, 3 – environments of the shallow shelf and coastal agitated shallow-water zone.

compaction of deep-water clays, recrystallization processes become the main compaction factor for clays deposited in other facies environments.

7.3.   MICROFRACTURING

The development of microfracturing is one of the complicated and poorly studied problems in the formation of clay rock properties. For a long time, the main attention in oil and gas geology was paid to the structural–tectonic factors which cause the break of continuity in the clay sequences playing the most important role in the formation and localization of fields. However, the study of macrofracturing does not solve the problem of clay seals completely. It is quite obvious that estimation of the clay seal quality within the undisturbed blocks of sedimentary rocks is impossible without taking into account their microfracturing.

The formation of microfractures is extremely complex and determined by the influence of both external (the stressed state of sequences, temperature) and a number of internal factors (such as the composition of rocks, homogeneity of their microstructure, character of structural bonds, presence of bound water, etc.). The least studied up to date is the role of rock's own peculiarities in the formation of microfractures. These peculiarities, in their turn, are connected with the conditions controlling the formation of clay sediments and their transformation in the course of lithogenesis. The following consideration of the problem will be focused just on these factors.

7.3.1.   GENERAL REGULARITIES

The main cause of microfracture formation is the concentration of stresses at local points (areas) of the rock and the irreversible break of its structural continuity. The conditions of stress concentration are determined by the character of structural bonds. No microfractures form in clays, where coagulation contacts between structural elements prevail. This is explained by two circumstances. The first one is that there are no conditions in such systems for the concentration of significant stresses at local points by virtue of the easy mobility of the structure and its high capability of dissipating stresses at the expense of mutual displacement of structural elements. The other cause is the quick reversibility (recreation) of coagulation contacts after their destruction. So, even when deformation is relatively quick, the number of coagulation contacts in the shear plane changes insignificantly. It is, thanks to these peculiarities of the coagulation contacts, that clays in a wet condition are capable of showing typical plasticity and deforming within a practically unlimited loading range without the break of continuity.

An important conclusion following from the aforesaid is that microfractures cannot arise within the diagenesis zone and at the beginning of the early catagenesis substage, when coagulation contacts are predominant in clays of any facies origin. The higher dispersion and wetting ability of minerals, the presence of Na^+ in the exchange complex, and the modification of mineral surface by humified organic matter favor the preservation of coagulation contacts in sediments and thus prevent the rise of microfractures also at the early and later substages of catagenesis.

The conditions for the formation of microfractures arise only when the transitional and especially phase contacts begin to prevail in clay sediments. But even the phase contact formation does not always lead to the rise of microfractures. The development of this process depends, besides the strength of contacts, on the homogeneity of clay microstructure, the rate of stress growth, and the pressure–temperature conditions. The presence in rocks of sand and silt grains, as well as of large micropores and microaggregates characteristic of sediments composed of illite, kaolinite, and chlorite, enhances the heterogeneity of microstructure and the probability of stress concentrator formation in it, which assists in the development of microfractures. The more homogeneous the microstructure and the less the sizes of structural elements composing it, the higher the probability that even in the presence of strong contacts the rock compaction will have the plasticized (creep) character. This is confirmed by field observations made by a number of researchers. I.I. Nesterov [172], for example, has noticed that an admixture of sand-silt and carbonate material in clays enhances their microfracturing.

The rate of loading and the character of the stressed state of sediments are of considerable importance. At a low rate of sediment accumulation, which takes place in the deep-water zone of a basin, the conditions for stress dissipation in the sediment structure are more favorable than in the shallow-water or middle-shelf environments, where the rates of sediment accumulation are higher.

Tectonic stresses and geodynamic processes contribute much to the rise of irregular stressed state in the rock structure. The formation of local platform- or transitional-type uplifts on the bottom of a basin may provoke the rise of microfracture zones in a clay sequence. In this case, as has been shown by A.M. Monyushko [229], the maximal microfracturing is observed on the periclines and flanks of uplifts, where rocks are subjected to the most intense deformation and extensive stresses.

When microfractures form under the effect of tectonic stresses, the rocks embedding the clay sequence play an important role. It has been noticed [230] that cemented rocks embedded in less rigid rock varieties show lower microfracturing than similar rocks embedded in very rigid varieties. This regularity is based on purely rheological phenomena. The less rigid rocks (e.g. salt) surrounding a clay sequence, by virtue of their plasticity, are easily deformed under the effect of tectonic stresses and transmit stresses uniformly to the more rigid clay rock sequence, where these stresses become gradually dissipated due to rheological displacements. But if clay rocks are enclosed in more rigid varieties, such as limestones or sandstones, the stresses are transmitted unevenly: the more rigid embedding rocks concentrate external stresses within themselves and only after their ultimate strength is reached, they quickly transmit the accumulated stresses to the clay rock sequence causing its brittle destruction rather than plasticized deformation.

The plasticized deformation of rocks due to rheological processes depends not only on the homogeneity of microstructure and rate of stress transmission, but also on the pressure–temperature conditions. With the growth of pressure and temperature, the probability of such a compaction mechanism increases, but only up to a certain depth of burial corresponding to the isothermal boundary of 65–70 °C. After this boundary is reached, the thermodynamic process of adsorption-to-free water transition begins, which is accompanied by a sharp increase in the strength of rock and in its brittleness.

The increase in the rock strength with the loss of adsorptionally bound water is explained by the effect of adsorptional strength decrease in solid surface wetting, which has been discovered for the first time by P.A. Rebinder.

In accordance with this effect, dehydration causes an increase in the surface energy of a body and, hence, also in its strength. A confirmation to it may be

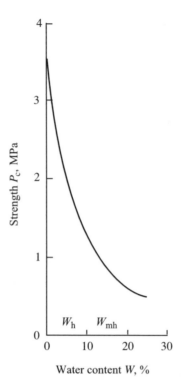

Fig. 7.11. Change in the strength of dry clay with the adsorption of moisture from the air.

the experimental data obtained by V.N. Sokolov [55] in studying a change in the strength of dry clay as a result of moisture adsorption by it from the air (Fig. 7.11).

With the loss of adsorptionally bound water and increase in the rock strength, the progressive process of microfracture formation starts, although its development may be different in different rocks. In fine grained, highly hydrophilic varieties, in which the squeezing-out of the formed free water is much impeded, this process may stretch over a great depth interval up to 6000 m and more. It is obvious that this is also the cause of slow development in such rocks of the hydromicatization of montmorillonite and mixed-layer varieties, accompanied by dehydration of interlayer intervals. In coarser and less hydrophilic rocks, the process of dehydration and development of microfractures may proceed more actively and be manifested at depths of 1800–3500 m, i.e. at the end of the middle catagenesis substage.

The development of microfractures at the above-indicated and greater depths may be enhanced due to the argillization of rock and increase in its schistosity. This process leads to the formation of large pseudocrystals of platy shape and to the increase of structural heterogeneity. Concurrently, the unevenness in the distribution of structural bonds increases and the conditions arise for their breaking along cleavage planes.

7.3.2. THE INFLUENCE OF DEPOSITIONAL ENVIRONMENTS AND LITHOGENETIC PROCESSES ON THE MICROFRACTURING OF CLAY SEALS

Consideration of general regularities in the formation of microfractures evidences for a close connection of this process with the composition, microstructure, and degree of lithification of clay rocks. Therefore, the regularities in the development of microfractures are peculiar for each facies type of seals.

Clay deposits of the deep-water shelf and adjacent sea environment, forming the seals of types I and II, are characterized, as has been already noted, by the high dispersion and wetting ability of constituent minerals (montmorillonite and mixed-layer varieties with high contents of montmorillonite layers), homogeneous microstructure, prevalence of Na^+ cation in the exchange complex. Thanks to these peculiarities, the deep-water clays are capable of retaining the bound water and preserving the coagulation contacts between structural elements up to the end of the middle catagenesis substage. The loss of adsorptionally bound water by these rocks, which begins after the isothermal boundary at 65–70 °C is passed, develops slowly due to the impeded squeezing-out of the forming free water and so does not cause a sharp increase in the strength and brittleness of rocks. This allows the clays under consideration to keep typical plastic properties up to depths of 3500–4000 m.

At greater depths, the transitional and phase contacts of the crystallization type form gradually, with the concurrent reduction of rock plasticity. This is promoted by the process of illitization which, though weakly developed, results in some increase in the contents of non-swelling clay minerals and in the lowering of rock dispersion. However, thanks to the rather high preserved homogeneity of microstructure and significant role of pressure–temperature factors, the deformation of rock continues to have plasticized rheological character and is not accompanied by the breaking of its continuity.

So, the clay seals of types I and II may be considered as preserving their continuity and showing no tendency to form microfractures up to the ultimate depths (5000–6000 m), which are of practical interest today.

Clay sediments composing the seals of types III and IV form in the environments of middle shelf and peripheral part of delta. Their characteristic features are the predominance of illite and mixed-layered minerals, content of clastic sand-silt grains amounting to 10–20%, lesser homogeneity of microstructure, and presence of significant quantities of Ca^{2+} in the exchange complex.

The peculiarities of the composition and microstructure of relatively deep-water sediments permit them to preserve hydrate films at the contacts of structural elements and to remain in the plastic condition to depths of 1500–2000 m (the end of the early – beginning of the middle catagenesis substages). As the transitional and phase contacts begin to form, the probability of microfracture growth at individual microstructural heterogeneities increases. This process, however, has a restricted development, and the rock continues to be deformed mainly at the expense of framework creep.

With the beginning of rock dehydration and transition of adsorptionally bound water into free state (depths 2500–3000 m), the process of microfracture formation intensifies considerably and the rock begins to lose quickly its isolating properties.

To confirm the aforesaid, we have generalized available data on the microfracturing of III- and IV-type seals for various oil- and gas-bearing basins. As an indicator of the clay condition, the fracturing modulus was used, which was determined as the number of microfractures per one linear meter of a core (F_c, m^{-1}). The obtained data (Fig. 7.12) show that the appearance of microfractures in rocks formed in the medium water depth environments is registered at depths of 1500–2000 m. The wide scattering of the fracturing modulus values (from 3 to 12) is obviously connected with the random development of this process and low representativeness of the data obtained. As the depth of rock burial increases, the fracturing modulus grows and attains values of 12–14 1/m at a depth of 2500–3000 m.

The depth dependence of the fracturing modulus for the seals of classes III and IV is expressed by the following correlation equation:

$$F_c = -63.416 + 9.586 \times \ln h; \quad R = 0.73 \tag{7.4}$$

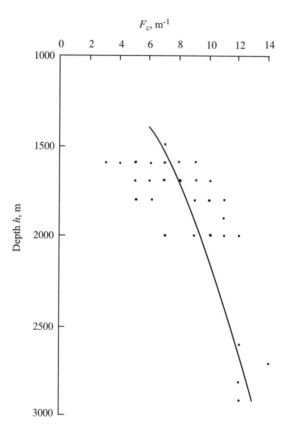

Fig. 7.12. Change in the microfracturing with depth for clay sediments of the facies of middle shelf and peripheral part of delta.

The highest tendency to develop microfractures is shown by clay rocks deposited in the shallow-water environment and composing the seals of types V, VI, and VII. Because of the weak wetting ability of principal rock-forming minerals, high heterogeneity of microstructure, significant contents of clastic grains (up to 70–80%), and active interaction with the outside solutions, the shallow-water clay rocks lose quickly their plasticity. Already at the early catagenesis substage, the process of cementation develops in them leading to the strengthening of rock and beginning of its destruction, which develops progressively with depth.

An analysis of microfracturing of the V–VII-type seals (Fig. 7.13) shows that the values of their fracturing modulus at depths of 1500–2000 m vary from 7 to 20, which is much higher than for the seals of types III and IV. At depths about 3500 m, the number of microfractures per 1 linear meter of core attains 35–40, which evidences for a high degree of rock destruction.

Processing of the obtained data on the fracturing modulus vs. depth dependence for the seals of types V–VII allows one to get the following equation:

$$F_c = -235.12 + 33.18 \times \ln h; \quad R = 0.76 \tag{7.5}$$

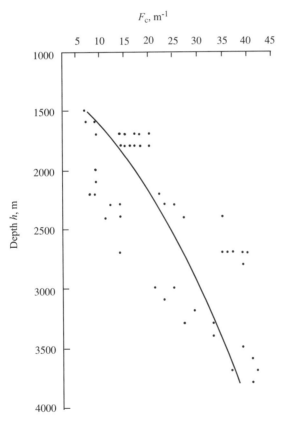

Fig. 7.13. Change in the microfracturing with depth for clay sediments of the facies of shallow shelf and coastal agitated shallow-water zone.

7.4.   PERMEABILITY

The main factors of clay rock permeability are their composition, porosity value and pore space structure, contents of bound water, microfracturing, salinity and chemical composition of filtering fluid, temperature and hydrodynamic gradients, homogeneity of the sequence composition along the section. The theoretical aspects of permeability and its dependence on various factors have been discussed in Chapter 3. So, below we shall analyze the influence exerted on the permeability by the composition, structure, and peculiarities of lithification of different facies types of clay seals.

7.4.1.   FACIES OF THE DEEP-WATER SHELF AND ADJACENT SEA BASIN

Sedimentation of clay material in the environments of a deep-water shelf and adjacent to it part of sea favors the formation of a sediment having a montmorillonite-mixed-layer composition with rather homogeneous small- and fine-pore microstructure. Already in the course of diagenesis, as a result of free pore water removal and closing of small micropores (1–10 μm), the sediment becomes compacted to the porosity value of 35–40%. The remaining in the sediments fine micropores (0.1–1 μm) and ultramicropores (<0.1 μm) prove to be obstructed to a large extent by adsorptionally bound and osmotic water, which reduces noticeably the effective porosity value and permeability of sediments. Further compaction of the clays at the early catagenesis substage results in the closure of larger quantities of fine micropores and practically complete cessation of filtration permeability. The ultramicropores remaining in the rock are completely obstructed by bound water and impermeable even at very high gradients. Thus, already at depths of 400–600 m, the permeability of deep-water sediments decreases to 10^{-5} mD, the sediments acquire good screening properties allowing them to become reliable seals for oil and gas fields.

Later, with further subsidence and compaction, the screening properties of deep-water clays continuously improve up to depths of 3500–5000 m. Insignificant modifications of the mineral composition occurring at these depths (illitization of mixed-layer varieties and montmorillonite) and transformation of adsorptionally bound water into free water do not substantially influence the screening properties because of the slow development of these processes. The absence of conditions for the development of microfractures allows the clay sequences of deep-water origin to remain impermeable for oil and gas down to the depths possibly exceeding 5000 m.

7.4.2.   FACIES OF THE MIDDLE SHELF AND PERIPHERAL PART OF DELTA

The sediments of this facies complex have a large-pore microstructure formed by the microaggregates of illite and mixed-layer minerals. As a result of compaction at the stage of diagenesis, permeability of these sediments drops quickly and by the end of the early catagenesis substage (depth 1200–1300 m) has the minimum value of 10^{-3}–10^{-4} mD, with rock porosity being 18–20%. The pore space is composed largely of fine micropores and ultramicropores filled to a considerable degree with bound water. Active porosity is formed at the expense of fine micropores and makes up an insignificant part of total porosity. With an increase in the hydrostatic pressure, the porosity grows a little. This is explained by the mobilization and inclusion in the filtration process of the external layers of bound water films.

It is of interest that total porosity of the facies under consideration is lower at depths of 1200–1500 m than that of the deep-water facies. In spite of this, their permeability has much higher values, which is caused by the pore space structure and by the presence in the sediments of the considered facies of fine pores not completely obstructed by bound water.

With the subsidence of sediments to greater depths into the middle catagenesis subzone, their filtering capacity, in spite of further compaction, does not decrease but begins to grow. This is connected with several causes. One of them is the development of microfractures caused by the cementation of clays and increase in their rigidity. The increase in the filtering capacity of the sequence is strongly influenced by the process of thermodynamic removal of adsorptionally bound water, which leads to the growth of effective porosity. Concurrently, the transformation of mixed-layer minerals into illite proceeds, which also promotes the increase in effective porosity. All these result in the noticeable growth of permeability, which at a depth of 2500–2800 m increases by an order of value and becomes equal to 10^{-3} mD.

With the beginning of argillization at the end of the middle catagenesis substage, the microfracturing increases and the rock becomes practically permeable not only to gas but also to oil.

Based on the available in literature data on the permeability of seals of types III and IV, a graph has been constructed by us, showing the change in permeability with depth (Fig. 7.14). As is seen from the given data, beginning with the depth of about 1500 m, the relatively deep-water clays start to increase progressively their permeability and at a depth of 2500–3000 m practically lose their screening properties.

The obtained dependence is expressed by the following relationship:

$$K = -0.168 + 0.023 \times \ln h; \quad R = 0.85 \qquad (7.6)$$

7.4.3. FACIES OF THE SHALLOW-WATER SHELF AND COASTAL ZONE

Clay sediments formed in the shallow-water environment have an irregular large-pore microstructure composed of low-activity non-swelling minerals and containing up to 80% of sand and silt grain inclusions. Despite the intense compaction of such a microstructure already at early stages of lithogenesis, it preserves its heterogeneity and relatively high active porosity. Therefore, the

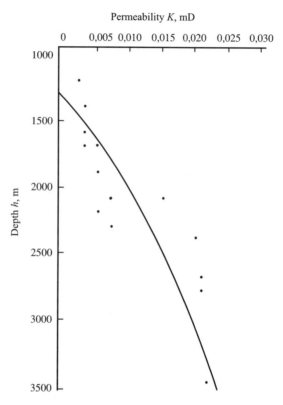

Fig. 7.14. Change in the permeability with depth for clay sediments of the facies of middle shelf and peripheral part of delta.

permeability decreases slowly and attains its minimal values $(10^{-2} mD)$ at the beginning of the middle catagenesis substage at depths of 1200–1800 m. The further increase in the screening properties of these clays stops because of their early cementation and development of microfractures.

Already at depths of about 1900–2100 m they cease to play the role of seals. After the thermodynamic removal of adsorptionally bound water, the clays lose completely their screening properties.

Few data are available in literature on the determination of filtration properties of shallow-water clay rocks. So, one may speak only of a certain trend in the change of K, mD, which consists in the gradual lowering of permeability beginning with a depth of 1500–1800 m.

7.5.   DESTRUCTION OF CLAY SEALS OF DIFFERENT FACIES TYPES WITH DEPTH

In the previous sections of the present chapter, the principal factors influencing the screening properties of clays of different facies types and the regularities in the change of these properties with depth have been analyzed. Below, approximate schemes of destruction of different-type clay seals will be considered.

Two tendencies compete in the formation of screening properties of clay seal. The first one provides the increase in the screening properties of clay sequences and is connected with the gradual compaction of rocks in the course of lithogenesis, increase in the homogeneity of microstructure, reduction of their general and effective porosity. The second one tends to decrease the screening property of clays and consists in the destruction of rocks. It includes: the formation of microfractures with the development of cementation and thermodynamic dehydration; increase in the heterogeneity of microstructure due to the mineral transformations (illitization) of montmorillonite and mixed-layer minerals; partial recrystallization of the rock matter in the process of its argillization; increase in the contents of bi- and trivalent cations in the exchange complex; increase in the effective porosity at the background of a gradual decrease in the total porosity value due to the lowering of the wetting ability of minerals with depth and the loss by them of the envelopes of adsorptionally bound water; decrease in the fluid viscosity with the growth of temperature.

The development of the enumerated above processes is determined by the initial mineral composition and dispersion (grain size distribution) of clays, their primary microstructure, the composition of exchange cations and the content of organic matter, as well as by all specific conditions of lithogenesis at its different stages. Both of the above-mentioned tendencies manifest themselves in a different way in clays of different facies and have specific regularities for each type of clays.

The seals of types I and II have the best isolating properties. The sediments forming them have a homogeneous small- and fine-pore microstructure with insignificant content of clastic grains. Under the influence of at first physicochemical and then gravitational compaction, these sediments acquire quickly the high screening properties. Already at depths of 400–600 m (the beginning of the early catagenesis substage), with porosity of 30–35%, they become reliable screens for oil and gas fields. This is explained by the practically complete loss by them of effective porosity at these depths owing to the obstruction of the majority of fine micropores and ultramicropores by bound water not participating in the filtration transfer of fluids.

At the following substages of catagenesis, their permeability continues to diminish due to further decrease in the total and effective porosity. Even the processes of mineral transformations and thermodynamic removal of adsorptionally bound water, which, though not actively, begin to develop in these rocks at depths of 2500 m and more, do not disturb the homogeneity of the rock microstructure and do not influence noticeably the permeability.

The extremely high dispersion and homogeneity of the microstructure of clays allows them at first to preserve true plasticity (at the expense of displacement of structural elements along hydrate films) and then, at greater depths, to show the plasticized deformation (creep) of the structure without breaking its continuity. This secures the occurrence of the sequence in an undisturbed condition up to depths of 5000–6000 m and possibly more.

So, the seals of types I and II represent good screens for gas and oil accumulations within an extremely wide depth range: from 400–600 m to the ultimate depths (5000–6000 m) of practical interest.

The seals of types III and IV are ranked among the moderately good screens: they are reliable in retaining oil pools and in most cases little permeable to hydrocarbon gases. The initial sediment has a large-cell microstructure formed by the microaggregates of illite and mixed-layer minerals. In the zone of diagenesis and at the early substage of catagenesis, the sediment undergoes intense compaction, whose rates exceed the rate of deep-water sediment compaction. Owing to the heterogeneity of microstructure and relatively low wetting ability of minerals, the rock still preserves the effective porosity, although its value diminishes more quickly than that of total porosity. At depths of 1000–2000 m, with porosity of 20–25%, the clays reach the best screening properties. Their permeability index decreases to 10^{-3}–10^{-4} mD due to the practically complete obstruction of active pore space by bound water.

The further compaction of rocks leads to the beginning of their cementation and increase in the strength and brittleness of structure. For some time, the strengthening rock deforms in a plasticized way at the expense of rheological processes, but already from a depth of 1800–2300 m it begins to show a gradual growth of microfractures and increase in permeability. When depths of 2500–3000 m are reached, the thermodynamic loss of adsorptionally bound water occurs and the rock loses completely its screening properties. This is promoted by the development of illitization characteristic of this type of sediments.

It follows from the above-said that the seals of types III and IV have the best isolating properties at depths of 1000–2500 m, with total porosity values of 12–25%.

The seals of types V–VII are poor screens for oil and are practically permeable to gas. The forming sedimentary sequence has a heterogeneous structure and contains interlayers and lenses of coarser material which drains well the water squeezed out of the sediment. A characteristic feature of the clay sediment is a heterogeneous large-cell microstructure with inclusions of silt-sand grains. The pore water is largely in free state.

With the growth of gravity forces, the sediment becomes quickly compacted, with the rate of its compaction exceeding that of the relatively deep-water and deep-water sediments. At the beginning of the middle catagenesis substage, with porosity of 15–20%, the rock acquires the maximal screening properties. In spite of considerable compaction, active pore space is preserved in the sediments and their permeability, therefore, does not decrease to less than 10^{-2} mD. The early cementation of the rock and the preserved heterogeneous microstructure cause the beginning of microfracture formation at relatively shallow depths not exceeding 1800 m. The further progressive growth of microfracturing quickly brings these rocks to the complete loss of screening properties.

It follows from the aforesaid that the clay seals of types V–VII formed in the shallow-water environments preserve their best screening characteristics within quite a small depth interval from 1200 m to 1800 m.

To conclude the consideration of the problem concerning the change in the condition and properties of different facies types of clay seals in lithogenesis, it is necessary to state a number of fundamental principles that follow from the analysis conducted:

1. The screening properties of seals are determined by depositional environments controlling the accumulation of clay material and increase with the depth of its deposition: the clays formed in the shallow shelf and coastal shallow-water environments have the lowest screening properties and those formed in the deep-water shelf and adjacent sea environments have the highest properties.

2. The depths at which the seals acquire isolating properties in the course of lithogenesis are the lesser, the higher the dispersion and homogeneity of sediment. Thus, clay sediments of the deep-water facies show high screening properties at depths of 400–600 m, those of the middle-shelf and peripheral part of delta facies – at 1000–1200 m, and those of the shallow shelf and coastal shallow-water facies – at 1200–1400 m.

3. The porosity of clay sediments, at which the best screening properties are reached, is the lesser, the lower the dispersion and homogeneity of a clay sequence: for the deep-water facies it corresponds to values of 30–35%, for sediments of the relatively deep-water facies it is 20–25%, and for sediments of the shallow-water facies – 15–20%.

4. The higher the dispersion and homogeneity of a clay sequence, the greater the depth range within which it preserves its best screening properties. The clay seals formed in the deep-water depositional environments preserve high screening properties within a depth range from 400–600 m up to 5000–6000 m and possibly to greater depths. The clay sequences formed in the middle-shelf and peripheral part of delta environments preserve them in the depth range from 1000–1200 m to 2000–2500 m. For the clay seals of shallow-water origin, preservation of the best screening properties is observed in a narrow range of depths measured by several hundreds of meters (1200–1800 m).

Part III

MODELING OF THE CHANGE IN THE CLAY SEAL PROPERTIES IN LITHOGENESIS

8

Physicochemical Models of Clay Rock Microstructures

8.1. THE CONCEPTS OF THE PHYSICOCHEMICAL MECHANICS OF CLAY ROCKS AS A BASIS FOR MODELING

Proceeding from the concepts of physicochemical mechanics, the structural-mechanical parameters of clay rocks depend not so much on the mechanical properties of the substance composing particles proper, but rather on the force of cohesion between solid structural elements and the number of contacts in the failure section. Therefore, the strength of fine grained structure P_s in simultaneous destruction of contacts is proportional to the average force of cohesion between particles P_1 and the number of destroyed contacts in the unit area of failure surface χ [17, 231, 232]:

$$P_s = P_1 \times \chi \qquad (8.1)$$

This dependence establishes a relation between the classical concepts about the strength of fine grained systems and its microstructural features.

In terms of physicochemical mechanics, clay rocks are heterogeneous multi-component systems with highly developed internal interphase surface, i.e. with excessive free interphase energy σ. The portion of the compensated interphase energy ΔF_a determined by the contact interactions between particles relative to the total excessive energy of the system is approximately the same as that of the contact area relative to the total interphase surface. It is just the value of ΔF_a that determines the capability of the system to resist external effects, i.e. it determines its mechanical strength. As a result of high open porosity of fine grained systems, the free energy of particle interaction ΔF_a in them is more sensible, in comparison with continuous bodies, to the influence of such physicochemical factors as water content, composition and concentration of pore solution, temperature, geostatic pressure.

The use of the described approach, which has found its application in studying the strength of fine grained systems [51, 54, 69, 233] allows one to approach the modeling of change in the properties of clay seals in the process of lithogenesis.

8.2.   CALCULATION OF THE NUMBER OF CONTACTS

An analysis of relation (8.1) shows that the strength values of rock are determined by its microscopic parameters (the strength of cohesion in a contact and the number of contacts in the unit area of failure surface), which in their turn depend on the micromorphological features of the structure. The more accurate these parameters will be determined and the more closely the calculation scheme reflects the real structure of a fine grained system (in the present case, this is the real microstructure of some or other type of clay seal), the more reliable the prognostic estimates for the behavior of the whole system under the influence of various physicochemical factors will be.

To evaluate the number of contacts in the failure plane, several calculation schemes-models developed for a fine grained porous body are presently available. The number of contacts in these models is determined by the size and shape of particles, mode of their packing and is closely associated with porosity. The less the size of particles and the closer their packing, the more contacts are contained in a unit volume and the higher the strength of the system is.

The simplest is the *globular model* proposed by P.A. Rebinder et al. [231] for the structures composed of spherical particles and having porosity more than 48%. Later, this model has been extended to include the structures with porosity ranging from 48 to 26% [234]. The globular model is constructed using rectilinear chains, which consist of spheres with equal diameters in contact with one another (Fig. 8.1). The chains are arranged along three mutually perpendicular directions and form the nodes of the structure at the intersections. The mode of packing is characterized by a structural parameter N – the average number of particles from node to node. If $N = 1$, the structure will have simple cubic packing, and if the value N is fractional, the system is characterized by irregular alternation of nodes. In such a model, porosity n is definitely related to the parameter N. This relation is usually given graphically

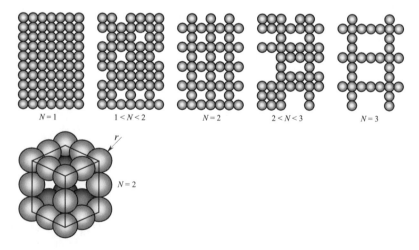

Fig. 8.1. A scheme of the globular model of fine grained porous structure.

(Fig. 8.2) as the following function [234]:

$$1/N^2 = f(n) \qquad (8.2)$$

The number of contacts per unit of failure surface is related to the parameter N and the average radius of a structural element r as follows [42, 231, 234]:

$$\chi = \frac{1}{4r^2 N^2} \qquad (8.3)$$

V.G. Babak [235] has proposed a different formula for finding the number of contacts in the globular model:

$$\chi = \frac{3}{2\pi} \frac{z(1-n)}{(2r)^2}, \qquad (8.4)$$

where z = coordination number, characterized by the average number of contacts of each structural element with the neighboring ones, n = porosity, r = average radius of a structural element.

The coordination number does not depend on the size of structural elements in the globular model and is determined solely by its porosity. To find z, there is a number of simple expressions derived by different authors. Thus, W.G. Field [236] has proposed the following relation:

$$z = \frac{12}{1+e}, \qquad (8.5)$$

where e = void ratio.

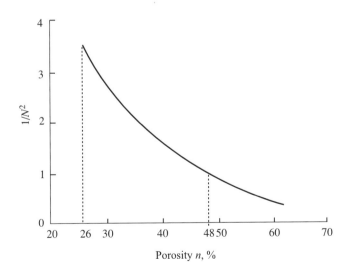

Fig. 8.2. Dependence of the parameter $1/N^2$ on the porosity n of the structure.

W.A. Gray [237] have obtained a number of other relations:

$$z = 3.1/n \qquad (8.6)$$

$$z = 2 \times \exp(2.4n) \qquad (8.7)$$

Porosity in Formulae (8.6) and (8.7) is expressed in fractions of a unite.

The globular model can be used for sands, sandstones, siltstones, and some fine grained rocks, whose structural elements (particles and microaggregates) have a shape close to a sphere. Thus for example, its application is possible for the calculation of the contact number in some loams composed of sand grains and rounded sand-clay aggregates (Fig. 8.3, a, b), as well as in tripolis, opokas, and opoka-like clays (Fig. 8.3, c) made up of silica globules. The calculations

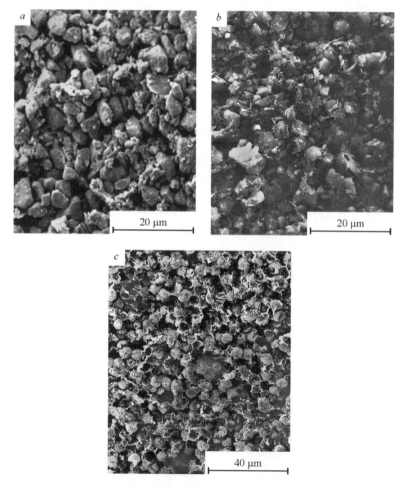

Fig. 8.3. Microstructure of clay rocks described by the globular model: a – loess-like loam; b – light marine loam; c – opoka-like clay.

performed have shown that that the number of contacts in opokas, depending on their porosity, is $10^5 \div 10^7 \, \text{cm}^{-2}$, and in loess-like loams it is 1.2×10^6 to $6.2 \times 10^7 \, \text{cm}^{-2}$.

The main shortcoming of the globular model is that it does not take into account the polydispersion and anisotropy in shape of structural elements composing clays. Therefore, V.N. Sokolov [238] has proposed a *bi-dispersed globular model*, which can be used for estimating the number of contacts in a system composed of large (with radius R) and small (with radius r) particles (Fig. 8.4). According to this model, the total number of contacts is equal to the number of contacts between large particles (χ_R) multiplied by the number of contacts between small particles (χ_r) which are present within the contact area between large particles. The values χ_R and χ_r are found from the following relations, respectively:

$$\chi_R = \frac{3z(1-n)}{8\pi R^2} \qquad (8.8)$$

$$\chi_r = \frac{\rho_R \varphi_r R^2}{2\rho_r \varphi_R r^2} \qquad (8.9)$$

where ρ_R, ρ_r, φ_R, φ_r, R and r are the density, content, and average equivalent radius of large and small particles, respectively; $z =$ coordination number, $n =$ porosity.

The total number of contacts in a bi-dispersed system is found as follows:

$$\chi = \chi_R \chi_r = \frac{3z(1-n)\rho_R \varphi_r}{16\pi r^2 \rho_r \varphi_R} \qquad (8.10)$$

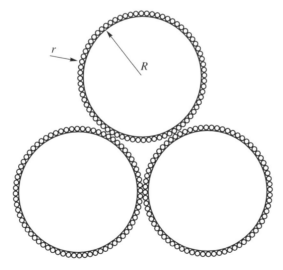

Fig. 8.4. A scheme of the bi-dispersed model of fine grained porous structure.

Fig. 8.5. Microstructure of clay rocks described by the bi-dispersed model: a – medium marine loam; b – light marine loam.

Calculations according to Formula (8.10) for loams (Fig. 8.5) show that the value χ for these samples with porosity of 40% may vary from $1.2 \times 10^7 \, \text{cm}^{-2}$ to $1 \times 10^9 \, \text{cm}^{-2}$ depending on the ratio of φ_R and φ_r.

For description of clay rocks composed of anisometric particles having a flat shape, a «*twisting card house*» model has been worked out [233, 238]. The basis of the model is the concept of a house of cards, whose walls are thin disks imitating platy clay particles or sheet-like microaggregates contacting with one another (Fig. 8.6).

To estimate the degree of rock compaction, parameter θ – the average value of angle between particles is used in the proposed model. Porosity is related to the θ value as follows:

$$n = 1 - \frac{K \times b/a}{\sin \theta + (K \times b/a)} \tag{8.11}$$

where K = coefficient (for disks, $K = 3\pi/4$), a and b = average values of disk diameters and thicknesses, respectively.

Based on the considered model, it is possible to estimate the number of contacts in a unit of failure surface (χ^θ) related to the average angle between clay particles (θ):

$$\chi^\theta = \chi^{90} \sin \theta \tag{8.12}$$

The number of contacts χ^{90}, when particles forming a cell are mutually perpendicular ($\theta = 90°$), is determined as $\chi^{90} = 2/a^2$, where a is the length of particle.

Taking into account the aforesaid, Formula (8.12) may be written as follows:

$$\chi^\theta = 2/\sin \theta \times a^2 \tag{8.13}$$

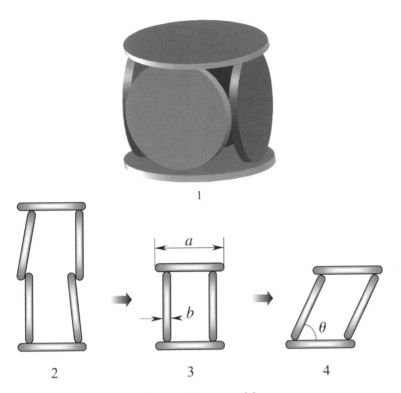

Fig. 8.6. A scheme of the «twisting card house» model.

The parameters a and θ may be estimated on the basis of the quantitative image analysis of clay rock structure obtained under scanning electron microscope.

The calculations accomplished according to Formula (8.13) have shown that the value of χ^θ for the present-day mud from the Caspian Sea (Fig. 8.7, a) is $6.9 \times 10^7 \mathrm{cm}^{-2}$. At the same time, for a marine Khvalyninan clay recovered in the area of Volgograd (Fig. 8.7, b) and having approximately the same grain size as the mud, but a higher degree of structural element orientation, the value of χ^θ equals to $3.9 \times 10^8 \mathrm{cm}^{-2}$. For a clay schist (Fig. 8.7, c), χ^θ increases to $4.4 \times 10^8 \mathrm{cm}^{-2}$. Not so great an increase of χ^θ occurs in this case because, in spite of the minimal value of angle θ, a characteristic feature of clay schists is the increase in the size of microaggregates (increase of the parameter a in Formula (8.13)).

Consideration of the described models for fine grained porous bodies has shown that some of them may be successfully used for the calculation of the number of contacts in clay seals. In this calculations, for the clay seals of classes I–II and III–IV represented mainly by clays of various lithification degree, argillites, and clay schists, the best results are obtained when the twisting card house model is used. This model takes into account all the characteristic features in the structure of such clay rocks: the anisometry of the sheet-like clay particle

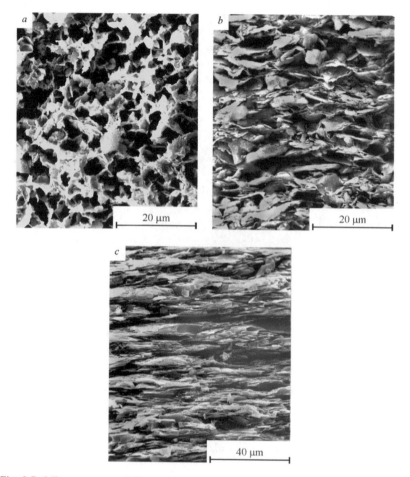

Fig. 8.7. Microstructure of clay rocks described by the twisting card house model: *a* – mud from the Caspian Sea; *b* – marine Khvalynian clay; *c* – clay schist.

microaggregates composing the rock, degree of orientation of the contacting structural elements.

For the clay seals of classes V–VII represented mainly by silty clays and siltstones, it is more expedient to use the bi-dispersed model. It takes into account that silty clays and siltstones are composed of predominantly rounded in shape grains of various minerals up to 30–50 μm in size covered by a continuous layer of clay particles, whose size does not usually exceed one micron.

8.3.   STRENGTH OF INDIVIDUAL CONTACTS BETWEEN CLAY PARTICLES

Calculation method.   The strength of an individual contact is the most important characteristic of clay rock structure. To estimate it, a method is often used based on the application of Equation (8.1) describing the relationship between the tensile strength P_t of the structure, the number of contacts per unit area of the failure section χ, and the average strength of individual contacts P_1. Assuming that under the tensile test, all the contacts become destroyed simultaneously, one can obtain from Equation (8.1) the average strength of a single contact:

$$P_1 = P_c/\chi \qquad\qquad (8.14)$$

Calculations accomplished by V.I. Osipov et al. [69] for clays with various degrees of lithification have shown that the value of P_1 varies within a significant range. Thus for example, for present-day muds, the strength value of individual contacts has been found to be ~10^{-10}N to 10^{-8}N, which corresponds to the far and close coagulation contacts. For weakly lithified varved clays, the P_1 values are ~10^{-8} to 10^{-7}N, which is characteristic of the close coagulation and transitional contacts. For argillites, values of ~6×10^{-7}N have been obtained providing evidence for the prevalence of phase contacts.

Experimental methods. Direct measurements of the strength of individual contacts are a complicated experimental problem requiring high-precision tests.

Coagulation contacts. The first direct measurements of cohesion forces between solids were the experiments carried out by B.V. Deryagin and I.I. Abrikosova [62] using a microbalance with negative feedback. It was for the first time that in the course of these experiments the value of molecular interaction between a quartz plate and a lens has been measured, dependence of the energy (force) of molecular interaction on the distance has been obtained, and Hamaker's constant A has been calculated to equal ~5×10^{-21}J for quartz. An analysis of the results of these experiments has shown that the force of molecular attraction between macrobodies (the quarts plate 3×7mm^2 in area and the quartz spherical lens with $R = 26$cm) at distances around 100 nm may attain 10^{-8}N.

The further development and perfecting of the theoretical and experimental basis have resulted in the creation of more sensitive apparatus allowing one to determine more accurately the dependence of the energy of interaction between mica plates on the distance between them and to estimate the influence of various physicochemical factors on this interaction. Thus, J.N. Israelachvili and G.E. Adams [28] have worked out a device, in which samples were drawn together by means of a piezoelectric displacement transducer and the force of contact interaction between the samples was determined from the flexure of a plane spring – dynamometer – with a precision up to 10^{-7}N. The width of clearance between the mica plate surfaces was measured using the method of multiple interference of a light beam with a precision of 0.1–0.2 nm.

By the aid of the described device, R.M. Pashley and J.N. Israelachvili [239] have obtained new data on the dependence of the interaction energy of mica plates on the concentration and type of electrolyte. In particular, they have established that for distances of more than 5 nm, the experimental results are in agreement with the DLVO theory. At the same time, at nearer distances and with high concentrations of KCl electrolyte ($C \geqslant 10^{-2}$mol/l), the forces of repulsion were observed, whose manifestation was attributed to the processes of hydration of potassium cations adsorbed on the mica surface.

At present, it has become possible to determine the interaction force between mineral particles using the atomic force microscopy (AFM).

The AFM method is comprehensively described in the following works [240–243]. It is based on the precision measurement of attractive force between a solid particle fixed on AFM cantilever and the particles at the surface of analyzed sample. Cantilever is moved be means of piezoelectric cell. The measurement accuracy of the clearance between two interacting particles ranges within 0.1–10 nm, the force sensitivity lying in the picoNewton range [244].

Application of AFM method permitted assessing forces between hydrophobic glass surface and quartz microsphere [245–247] and compiling a force curve for the hydrophobic particle interaction in aqueous medium [244].

Transitional and phase contacts. To measure the forces acting at the transitional (point) and phase contacts, a special device has been developed allowing the measurement of cohesion in contacts between two solid particles under different conditions of their pressing [234]. A high-sensitivity magnetoelectric system of galvanometer (Fig. 8.8) was used in the device as a loaded element and a force-measuring device. The force was measured with a precision of 10^{-8} N. In the course of experiment, two particles, one of which (*a*) is firmly fixed on the galvanometer pointer (1) and the other (*b*) – on the manipulator (2) are drawn together till they come into contact. The moment of the contact is registered by means of a long-focus microscope. The effort of particle pressing together and tearing off is created by passing the current through the galvanometer frame in one and the other direction. The force measured corresponds to the tensile

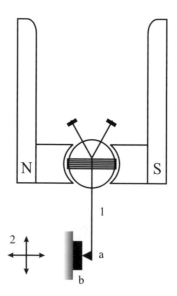

Fig. 8.8. A scheme of the device for measuring the force of cohesion in the contacts between solid particles.

strength of contact, i.e. reflects the total action of structural bonds in the contact. The device allows measurements in the air and in liquid media at temperatures from room temperature to 200 °C.

Experimental investigations conducted by R.K. Yusupov [111] using the above-described device have shown that at pressing forces $f_p > 5 \times 10^{-5}$N in microcrystals of different mineral composition (naphthalene, sodium chloride, talc, and vermiculite), plastic deformation of particles begins in the contact areas, resulting in the formation of at first transitional ($P_1 \approx 10^{-7}$N) and then phase ($P_1 > 10^{-7}$N) contacts. The presence of surface-active substances at the contacts produces a stabilizing action and may decrease the contact strength by several orders of values.

V.P. Vaganov [248] has investigated experimentally and described another mechanism of phase contact formation: at the expense of point cementation – a primary bridge connecting mineral particles, which arises under certain physicochemical conditions. Direct measurements performed using the above-considered magnetoelectric system have shown that phase cementation contacts with $P_1 \geqslant 10^{-6}$N form between crystals of dehydrous gypsum in oversaturated solutions of calcium sulfite.

In conducting the experiments on direct measurements of cohesion forces in contacts, the exact fixation of the moment when particles come into contact and determination of the contact area after its destruction are especially difficult. V.G. Babak et al. [249]; V.N. Sokolov [233] have worked out a special device allowing the investigation of contact interactions immediately in the specimen chamber of SEM. In determining the cohesion forces between particles or microcrystals, one sample is placed on a standard goniometric device of SEM and the other one is fixed on a holder rigidly connected with a magnetoelectric system, whose proper rigidity is 9×10^{-3}N/m. The smooth drawing together of the samples, their mutual pressing and following tearing-off were realized by means of passing the current through the frame of the system. The precision of measurement of mechanical efforts in the contact area of samples is 5×10^{-9}N, and the range of the measured forces is 10^{-3} to 10^{-8}N.

Figure 8.9, *a* is a photography showing the moment of contact between a NaCl microcrystal and mica surface coated with gold (at this moment, an increase in the pressing was registered), and Figure 8.9, *b* shows the imprint on the mica surface after the destruction of its contact.

Later, the precision of measurements of the force and clearance between the interacting particles was increased [64]. The developed device (Fig. 8.10) allows measurements of the pressing force f_p and cohesion force f_a between two crossed cylindrical surfaces of bent plates of mica in electrolyte solutions, as well as modeling the face-to-edge interactions between crossed plates of mica contacting according to the «plate edge – cylindrical surface» type.

The precision in measuring the clearance between particles with the use of the described device was 2 nm and the precision in determining the forces of contact interaction at $f_a = 10^{-7}$N was as high as 5×10^{-9}N.

Experimental investigations using the described device have allowed V.N. Sokolov [233] to estimate the value of the total effect of the molecular

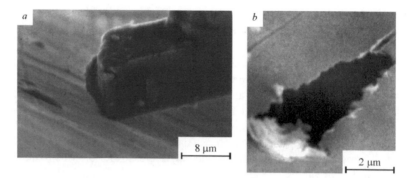

Fig. 8.9. SEM photographs: *a* – the moment of contact between NaCl microcrystal and mica surface; *b* – imprint of the contact on the mica surface.

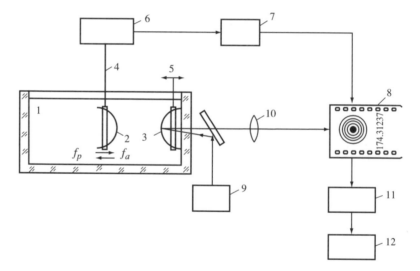

Fig. 8.10. A scheme of the device for investigating the contact interactions between mineral particles in various media: 1 – transparent quartz dish; 2, 3 – mobile and fixed specimens (mica sheets); 4 – holder; 5 – special specimen holder; 6 – magnetoelectric dynamometer; 7 – microamperemeter; 8 – interference pictures filmed with lensless motion-picture camera; 9 – helium-neon laser; 10 – microscope; 11 – image analyzer; 12 – personal computer.

and ionic-electrostatic forces acting between mineral particles of muscovite mica, as well as to calculate the ratio between the electrostatic interaction and the total molecular and ionic-electrostatic interactions, which is equal to 0.48. In addition to this, experimental data on the influence of physicochemical factors on the contact interactions of particles have been obtained.

Figure 8.11 presents the results of direct experimental investigations of the cohesion forces interacting according to the schemes of crossed cylinders and

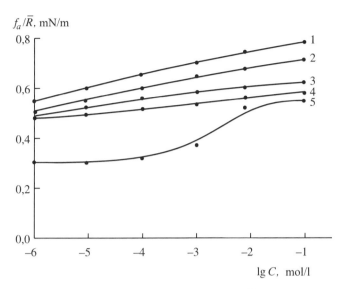

Fig. 8.11. Dependence of the reduced cohesion force $f_a/\bar{R}$ between plates of mica (muscovite) on the concentration C of KCl in electrolyte at the fixed reduced pressing force $f_p/\bar{R} = 10\,\text{mN/m}$: 1, 2 – interaction of crossed cylindrical surfaces of mica at pH = 3 and 12, respectively; 3–5 – the edge – cylindrical surface of mica interaction at pH = 3, 6, and 12, respectively.

plate edge – cylindrical surface in electrolyte solutions with different pH and salt concentrations. Experiments with crossed cylinders have shown that at pressing forces corresponding to the reduced forces of contact pressure between real clay particles ranging from 10^{-1} to $10^2\,\text{mN/m}$, the reduced force of cohesion $f_p/\bar{R}$ remains practically constant. This evidences that the interaction between mica surfaces occurs when they draw together as close as the near potential minimum. With the growth of KCl concentration in the electrolyte solution, an increase in the reduced cohesion force in the contact is observed, which is seen from the trends of curves 1 and 2 at the fixed pressing force $f_p/\bar{R} = 10\,\text{mN/m}$. The indicated dependence is in accordance with the conclusions of the DLVO theory concerning the influence of electrolyte on the depth of near potential minimum. A change in the pH of electrolyte exerts a weak influence on the force of cohesion of crossed cylindrical mica surfaces (curves 1 and 2), as no significant changes occur in this case in the structure of the interacting likely charged electrical double layers on the basal surfaces of mica.

The more intense influence of physicochemical factors is manifested when one studies the forces of cohesion between mica plates interacting according to the edge – surface of cylinder type (curves 3–5). At low values of pH (up to 6), the form of curves 3 and 4 is similar to that of curve 1 for the crossed cylinders. However, under alkaline conditions (pH = 12) at small concentrations of KCl, a sharp decrease is observed in the reduced force of cohesion, whose value grows as the electrolyte concentration increases to 1 mol/l (curve 5). The decrease in the reduced cohesion force at low values of electrolyte concentration in the alkaline medium is explained by the fact that the plate edges under these conditions change the positive charge to the negative one. This is accompanied by the rise of additional ionic-electrostatic repulsion forces of EDL with like charges between the edge and face parts of mica plates, which reduce the cohesion force on the whole. In the neutral and acid media, the edge part of mica plate is positively charged and the arising electrostatic attraction between the positive edge and the negative basal surface of mica plate increases the cohesion force. Addition of salt to the alkaline electrolyte results in the suppression of the ionic–electrostatic repulsion forces and compression of the diffuse layer, which causes an increase in the molecular attraction and in the cohesion force on the whole.

9

Development of a Mathematical Model for Estimating the Strength of Clay Seals

Development of physicochemical models of clay rock microstructure, as well as creation of the theory of contact formation between clay particles in the course of lithogenesis open new possibilities for obtaining information about the mechanical properties of clay rocks.

In connection with this, the authors have proposed a model for estimating the strength of rocks composing the clay seals, based on the use of Equation (8.1) relating the macroscopic strength of a sample P_s to the microscopic parameters of structure – χ (the number of contacts in a unit area of the failure section) and P_1 (the strength of an individual contact).

The essence of the method is that a fundamental theoretic dependence of the strength of individual contacts on the depth of clay rock occurrence $P_1 = f(h)$ may be obtained for clay seals of specific classes. This dependence forms the basis for estimating the strength of clay seals from the data of microstructural investigations.

Further, one can carry out a quantitative microstructural analysis, determine the required parameters of the physicochemical model (e.g. «a» and $\sin \theta$ for the twisting card house model), and calculate another important parameter of the model χ – the number of contacts in a unit area of the sample failure section. Then, knowing the depth of sample recovery, one can determine from the graph $P_1 = f(h)$ the strength of individual contacts and using Formula (8.1) calculate the tensile strength of a clay seal (P_t) and the uniaxial compression strength (P_c).

9.1. THEORETICAL CALCULATION OF THE CHANGE IN THE STRENGTH OF INDIVIDUAL CONTACTS WITH DEPTH

As has been noted earlier, a regular change in the type of contact occurs in the course of clay seal lithogenesis. At the initial stage of sediment accumulation (in sedimentogenesis), distant coagulation contacts are prevalent in clay sediment ($P_1 = 10^{-10} \div 10^{-9}$ N for the particles $d \approx 1\,\mu m$ in size), which, as the compaction increases at the stage of diagenesis, convert into the close coagulation contacts ($P_1 = 10^{-9} \div 5 \times 10^{-8}$ N). In the course of the following transformation of clay seals at the substages of early and middle catagenesis, transitional contacts ($P_1 = 5 \times 10^{-8} \div 3 \times 10^{-7}$ N) form in them, which in late catagenesis convert into the strongest type – the phase contacts ($P_1 > 3 \times 10^{-7}$ N). With further lithogenetic transformations of clay rocks at the stage of metagenesis, the phase contacts increase in area, which is accompanied by the further growth of individual contact strength ($P_1 = 10^{-6} \div 10^{-5}$ N).

The described scheme of contact transformation in clay rocks in the course of lithogenesis is a fundamental relation, however, real strength values of individual contacts of different types may vary in a wide range depending on the type of clay rock, its mineral composition, grain size distribution, degree of structural element orientation, character of pore space, composition and concentration of pore solution.

In accordance with the above-stated, it should be expected that the character of the relation between P_1 and the depth of rock occurrence would change significantly in clay seals depending on their composition and depositional environments. Given below is a theoretical calculation of the relation between the strength of individual contacts and the depth for clay seals of different classes.

9.1.1. DEPENDENCE OF THE STRENGTH OF INDIVIDUAL CONTACTS ON THE DEPTH OF ROCK OCCURRENCE FOR CLAY SEALS OF CLASSES I–II AND III–IV

An analysis of published data and the authors' own experimental investigations have shown that clay seals of classes I–II and III–IV are mainly represented by clays of different lithification degree, argillites, and clay schists (Fig. 9.1, *a, b*).

These rocks are usually composed of clay particles having anisometric platy shape and are characterized by the increase in the degree of orientation of structural elements with depth [23, 233].

Figure 9.2 shows the change in the angle of inclination between contacting clay particles with depth. These data have been obtained in the course of a quantitative microstructural analysis conducted on a large collection of clay rock samples representing clay seals of classes I–II and III–IV. The angle between clay particles was determined with an image processing system involving SEM combined with personal computer (PC). A specially worked out software «STIMAN» was used for this aim [250].

As it follows from the data obtained, the most significant restructuring of microstructure occurs in clay seals of classes III–IV (Fig. 9.2, curve 2) in the zones of diagenesis and early catagenesis extending to the depth of 1300 m. At these stages, the angle θ sharply diminishes from 80° to 23°, which is accounted for by the lower content in such rocks of swelling clay minerals and more quick removal from them of pore water as they become compacted with depth. This process is accompanied by the intense transformation of distant coagulation contacts into the close coagulation and then transitional contacts.

In the zones of middle and late catagenesis (depths $\sim$1300–3800 m), when transitional contacts become transformed into phase ones and mutual fixation of clay particles begins, the process of rock compaction slows down and further

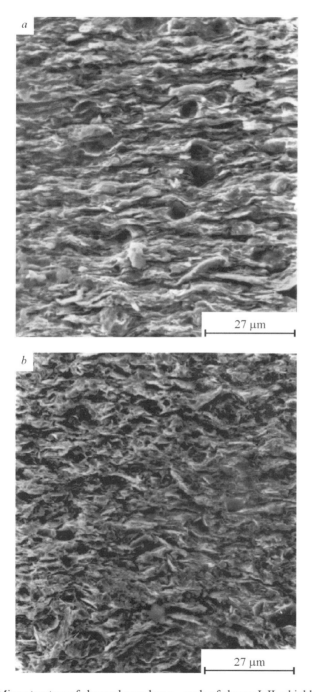

Fig. 9.1. Microstructure of clay seal samples: a – seals of classes I–II, a highly lithified clay, D_2, depth 4020 m, Tatarstan, Romashkino field; b – seals of classes III–IV, argillite, K_1, depth 2000 m, Krasnodar region.

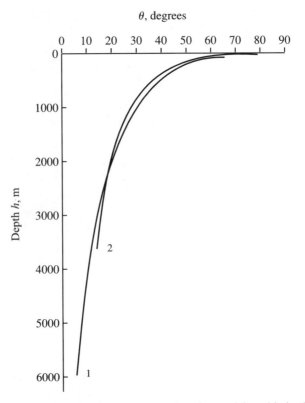

Fig. 9.2. Change in the angle θ between contacting clay particles with depth h: 1 – for clay seals of classes I–II; 2 – for clay seals of classes III–IV.

restructuring of microstructure becomes insignificant. At these stages, the angle between particles changes only from 23° to 12°.

For clay seals of classes I–II, a more smooth character of the angle θ change with depth is observed in comparison with the III–IV-class seals. It is seen from Figure 9.2, curve 1, that the most intense reorientation of clay particles in the seals of classes I–II is completed at the end of the early catagenesis stage at depths of ~1800 m, where the transformation of close coagulation contacts into transitional ones occurs. In this case, the angle θ diminishes from 80° to 22°.

At the stages of middle and late catagenesis, up to depths of ~6000 m, a slow decrease of angle θ from 22° to 5° is observed. Here, further transformation of close coagulation contacts into the transitional and phase ones occurs.

Such a character of compaction of I–II class clay seals is explained by the greater contents of swelling minerals in these clay rocks and by their small particle size. The indicated peculiarities of the mineral and granulometric composition impede the process of compaction of these rocks and the squeezing of pore moisture out of them, which may favor the conservation of some quantity of close coagulation contacts and ensure their presence at great depths. This

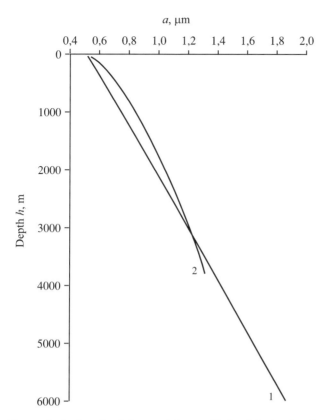

Fig. 9.3. Change in the length of clay particles *a* with depth *h*: 1 – for clay seals of classes I–II; 2 – for clay seals of classes III–IV.

fact is confirmed by discoveries at depths as great as 4000–6000 m of I–II class clay seals having a tight-plastic and semi-solid consistence.

An analysis of the granulometric composition of clay seals ranking with classes I–II and III–IV, as well as microstructural investigations in SEM have allowed the authors to estimate also the change in the size of clay particles (along the major axis) with the depth of rock occurrence. These data are given in Figure 9.3. It follows from them that, as clay rocks subside in the course of lithogenesis, the length «*a*» of contacting clay particles increases. Besides, the length of clay particles in clay seals of classes I–II increases more intensely from 0.5 μm at the 30 m depth of rock occurrence to 1.87 μm at their maximal depth of 5700 m (Fig. 9.2).

In the clay seals of classes III–IV, the growth of clay particle length is less intense. It changes from 0.52 μm at the depth of 30 m to 1.32 μm at 3800 m.

The growth of clay particle size with depth is explained by the specific change in the mineral composition of clay seals. As has been shown earlier (see Chapter 6), as the depth of clay seal occurrence grows, their contents of swelling minerals (montmorillonite, mixed-layer minerals) gradually decrease, while those of non-swelling minerals (hydromica, kaolinite, and chlorite) increase. Taking into account that clay particles of non-swelling minerals are larger in size than those of swelling minerals, their increase in content with depth can explain the observed growth in size of solid structural elements in clay seals down the section.

Calculation of the strength of individual contacts for seals of classes I–II and III–IV. The possibility of a quantitative estimation of parameters «*a*» – the length of clay particles and θ – the angle between contacting particles and of their change with depth at different stages of lithogenesis has allowed the authors to obtain a theoretical dependence of the strength of individual contact P_1 on the depth h of clay seal occurrence.

Taking into consideration the similar (platy) morphology of clay particles composing the clay seals of classes I–II and III–IV, theoretical calculation of the change in P_1 with depth was carried out for them according to one scheme.

For the stages of early and late diagenesis, when distant and close coagulation contacts between clay particles prevail in a clay sediment, the calculation of P_1 was carried out on the basis of an equation describing the balance of attraction and repulsion forces between hydrated clay particles. According to V.N. Sokolov [233], the force of cohesion between hydrated clay particles may be presented as follows:

$$f_a = f_m(\theta) + f_{i-e}(\theta) + f_e \qquad (9.1)$$

where $f_m(\theta)$ – molecular forces depending on the angle θ between particles and determined only by the properties of the particle material and medium of occurrence; $f_{i-e}(\theta)$ – ionic-electrostatic repulsion of diffuse layers with like charges on the basal surfaces of particles, depending on the angle θ between particles, clearance between particles, and electrical double layer thickness; f_e – electrostatic interaction between the charged edge and the basal face of clay particles, not depending on angle θ; f_e depends on the value and sign of charge on edges and basal surfaces: it may be attraction in acid and neutral media and repulsion in an alkaline medium [51, 59, 251, 252].

Taking into account the theoretical relations known for the energy of interaction of two clay particles in their turning through angle θ, the equation for f_a can be presented in the following form:

$$f_a = f_e + \frac{f_m^{90} + f_{i-e}^{90}}{\sin^{1.5}\theta} \qquad (9.2)$$

where f_m^{90} and f_{i-e}^{90} are force values at $\theta = 90°$.

V.N. Sokolov [233] has performed a calculation of the cohesion force between two kaolinite particles under neutral conditions in distilled water to obtain: $f_e = 1.2 \times 10^{-9}$ N, $f_m + f_{i-e} = 3.2 \times 10^{-9}$ N, and $f_a = 4.4 \times 10^{-9}$ N.

An important implication of Formula (9.2) is that with the decrease of angle θ, the cohesion force in the coagulation contact must increase owing to the growth of contact area.

Thus, knowing the relation describing the change of angle θ between contacting particles with depth and using Formula (9.2), the authors have calculated the change in the strength of individual coagulation-type contacts P_1 with depth h.

Seals of classes I–II.   As calculations for clay seals of classes I–II have shown, P_1 changes from 4.5×10^{-9} N at a depth of 30 m ($\theta \approx 80°$) to 2.78×10^{-8} N at depths around 300 m ($\theta \approx 48°$). These data indicate that at the beginning of the stage of late diagenesis, close coagulation contacts with $P_1 < 4.5 \times 10^{-9}$ N start to form and at the end of the stage of late diagenesis (at depths around 300 m) the close coagulation contacts with $P_1 \approx 2.78 \times 10^{-8}$ N prevail. Further calculation of P_1 is carried out proceeding from the fact found earlier by the authors, that transitional contacts begin to form at the stage of early diagenesis and then are gradually transformed into phase contacts in the middle and late catagenesis.

Taking into account that one of the possible mechanisms for the formation of transitional and phase contacts in clay rocks during their compaction and dehydration is the rise of cation «bridges» between clay particles [52, 54], the authors performed the estimation of P_1 for these contacts using the scheme for calculating the forces of ionic–electrostatic attraction.

To calculate the force of ionic–electrostatic attraction, let us consider a system consisting of two plane clay particles 1–2 with dielectric permittivity ε_m placed in water with variable dielectric permittivity ε_w (Fig. 9.4). The distance between the particles is $2d$ and a negative electric charge with density σ is distributed over their surface. Positively charged univalent exchange cations (3) with charge $+ q = e$ are located in the middle between the particles.

Two kinds of interactions may exist in such a system: attraction and repulsion. Attraction is made up of:

1. Attraction between the negatively charged face σ and positive cation

$$F_1 = \frac{2 \times \pi \times \sigma \times e}{\varepsilon_w} \tag{9.3}$$

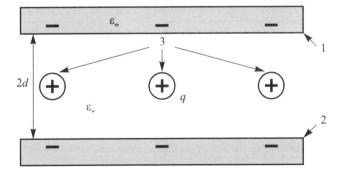

Fig. 9.4. A scheme of calculation of the ionic–electrostatic attraction between clay particles (1–2) due to exchange cations (3).

2. Attraction between the positively charged cation and its mirror reflection

$$F_2 = \frac{(\varepsilon_m - \varepsilon_w) \times e^2}{8\varepsilon_w^2 \times \varepsilon_m \times d^2} \tag{9.4}$$

Repulsion is made up of:

1. Repulsion between the basal faces of clay particles with like charges

$$F_3 = -\frac{2 \times \pi \times \sigma^2 \times S}{\varepsilon_w} \tag{9.5}$$

2. Repulsion between the charged basal face of clay particle and the mirror reflection of the charge

$$F_4 = -\frac{2 \times \pi \times (\varepsilon_m - \varepsilon_w) \times e \times \sigma}{\varepsilon_w^2 \times \varepsilon_2} \tag{9.6}$$

The summary force per one cation will be:

$$F = F_1 + F_2 - F_3 - F_4 \tag{9.7}$$

In their calculations, the authors estimated the maximal ionic–electrostatic force for clay in the air dry condition ($P/P_s \sim 0.4$) and so took $d = 4.5$ Å, with other values taken as follows: dielectric permittivity of bound water ε_w and clay mineral (montmorillonite) ε_m equal to 8 and 4, respectively [253, 254], surface charge density for montmorillonite $\sigma = 1.01 \times 10^{-5}$ coul/cm^2, area occupied by one cation $S_c = 80$ Å^2 [253], $q = 1.6 \times 10^{-19}$ coul. A calculation accomplished according to Formulae (9.3)–(9.7) has shown that the total force per one univalent cation is $F = 3 \times 10^{-11}$ N [54] and per a bivalent or trivalent cation it is 2 and 3 times as great, respectively [52, 54].

To estimate the strength of an individual contact, it is necessary to determine the number of single bonds N within the contact area. The parameter N may be found from the ratio of the contact area between particles S to the area occupied by one cation S_c:

$$N = S/S_c \tag{9.8}$$

The maximal number of bonds N will be for the parallel arrangement of particles, when the contact area S is equal to the particle area a^2 (a – linear size of a particle).

Taking into account that the contact area will change significantly depending on the distance between particles $2d$ and the angle θ between them, one can write:

$$S = \frac{a^2 \times 2d}{\operatorname{tg}\theta} \tag{9.9}$$

On the basis of Formulae (9.8) and (9.9), a formula for the calculation of individual contact strength can be derived:

$$P_1 = F \times N = F \times \frac{a^2 \times 2d}{\mathrm{tg}\,\theta \times S_k} \qquad (9.10)$$

It is seen from Formula (9.10) that depending on the change in parameters a, $2d$, and θ, the values of P_1 may vary within a wide range. A theoretical calculation of the change in the strength of transitional and phase contacts with depth for the seals of classes I–II has been performed beginning with the depth of 300 m, as in the overlying clays the coagulation contacts are dominant. In this calculation, the authors used the following value ranges: particle length a varied from ~0.6 μm at the depth of 300 m to ~1.8 μm at the depth of 5700 m; distance between particles $2d = 20 \div 10$ Å; angle between particles $\theta \approx 48 \div 5°$ (Figs. 9.2 and 9.3); the area occupied by one cation S_c was taken equal to ~80 Å^2 [253]. The calculations accomplished according to Formula (9.10) have shown that the strength of individual contacts changes within a depth range of 300–5700 m from ~2.8×10^{-8} N to ~9×10^{-7} N.

Figure 9.5 (curve 1) shows on a semi-logarithmic scale a theoretical curve of change in the individual contact strength P_1 with depth h. This relation is

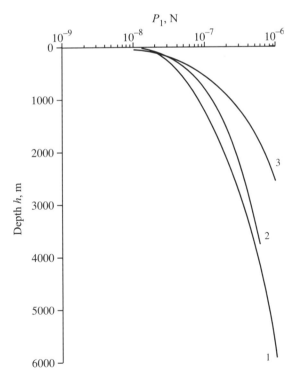

Fig. 9.5.   Theoretical curves of change in the strength of individual contacts P_1 with depth h for clay seals: 1 – of classes I–II; 2 – of classes III–IV; 3 – of classes V–VII.

described by a quadric polynomial equation of the following form:

$$P_1 = 1.41688 \times 10^{-8} + 4.16346 \times 10^{-11} \times h + 1.99155 \times 10^{-14} \times h^2$$

$$(9.11)$$

The high values of the correlation coefficient $R = 0.99$ indicate very close interrelation between the parameters P_1 and h.

It is seen from this graph that the most intense growth of P_1 occurs up to depths of ~300 m, i.e. at the stages of early and late diagenesis. At these stages, close coagulation contacts with $P_1 \leqslant 2.7 \times 10^{-8}$ N prevail in clay sediments. As the depth of clay rock occurrence increases to ~1800 m, the growth of P_1 slows down, and by the end of the early catagenesis stage $P_1 \approx 1.6 \times 10^{-7}$ N. These data show that at the end of the stage of early catagenesis, the close coagulation contacts become transformed into transitional ones.

When clay rocks subside to depths of ~3600 m, the intensity of P_1 growth decreases noticeably and it attains ~4.2×10^{-7} N. This evidences that by the end of the middle catagenesis stage, the almost all of close coagulation contacts in clay seals of classes I–II have transformed into transitional ones and that the transformation of transitional contacts into phase ones is going on.

And at last, at depths greater than 3600 m, up to the maximal depth of ~5700 m, a rather slow increase in the strength of individual contacts to the value of ~9×10^{-7} N is observed. Thus, at the stage of late catagenesis, further transformation of transitional contacts into phase ones occurs. Such a trend of curve $P_1 = f(h)$ and the values of P_1 at depths over 3600 m obviously do not exclude the presence in the clay seals of classes I–II of a certain quantity of transitional contacts also at greater depths (~4000–6000 m), as has been already noted before.

Seals of classes III–IV. Similar calculations have also been accomplished for the clay seals of classes III–IV. They have shown that in this case, taking into account the character of change in the particle size a and angle θ between particles depending on the depth of occurrence (Fig. 9.2, curve 2, and Fig. 9.3, curve 2), the process of increasing of the distant coagulation contacts strength from $P_1 = 6.2 \times 10^{-9}$ N (at $h = 30$ m) up to $P_1 = 2.11 \times 10^{-8}$ N (at $h = 150$ m) occurs at shallower depths, up to 150 m. Thus, such a transformation of contacts occurs chiefly at the stage of late diagenesis.

A theoretical calculation of P_1 for transitional and phase contacts was performed, as in the case of clay seals of classes I–II, according to Formula (9.10).

In this case, the values of parameters a, $2d$, and θ at depths of 150–3800 m varied within the following ranges: $a \approx 0.5 \div 1.3$ μm, $2d \approx 20 \div 10$ Å, $\theta \approx 38 \div 12°$.

The accomplished calculations have shown that the strength of individual contacts in the depth interval from 150 to 3800 m may vary from ~2.1×10^{-8} to 5.7×10^{-7} N.

Figure 9.5 (curve 2) shows a theoretical dependence of the strength of individual contacts on the depth of occurrence for clay seals of classes III–IV. This dependence is described by a quadric polynomial equation of the following form:

$$P_1 = 9.32472 \times 10^{-9} + 1.00533 \times 10^{-10} \times h + 1.21648 \times 10^{-14} \times h^2$$

$$(9.12)$$

The high values of the correlation coefficient $R = 0.99$ confirm the presence of very close interrelation between the parameters P_1 and h.

It follows from this graph that the most intense growth of individual contact strength occurs up to a depth of ~ 150 m, i.e. at the stages of early and late diagenesis, with distant coagulation contacts being transformed into the close ones with $P_1 \approx 2.1 \times 10^{-8}$ N.

As the depth h increases to ~ 1300 m, the growth of P_1 somewhat slows down and attains the values of $\sim 1.6 \times 10^{-7}$ N. This indicates that during early catagenesis, the majority of close coagulation contacts become transformed into the transitional ones. At the stage of middle catagenesis, when clay rocks of this type subside to depths of ~ 2800 m, the growth of curve $P_c = f(h)$ slows down significantly and attains $\sim 3.9 \times 10^{-7}$ N. Thus, at this stage of lithogenesis, the transformation of the remaining part of close coagulation contacts into the transitional contacts continues and the transformation of transitional contacts into the phase ones begins.

At greater depths, up to 3800 m, the slowest growth of P_1 to a value of $\sim 5.7 \times 10^{-7}$ N is observed. All this allows one to maintain that during late catagenesis, complete transformation of transitional contacts into phase ones occurs in clay seals of classes III–IV. This is also confirmed by the fact that samples of clay seals of classes III–IV (argillites and clay schists) recovered from depths of 3000–3800 m are characterized by high strength and do not get soaked in water.

9.1.2. DEPENDENCE OF THE STRENGTH OF INDIVIDUAL CONTACTS ON THE DEPTH OF ROCK OCCURRENCE FOR CLAY SEALS OF CLASSES V–VII

In contrast to the clay seals of classes I–II and III–IV, clay rocks composing the seals of classes V–VII have a coarser granulometric composition and are represented mainly by silty clays and siltstones. They are made up of large silt grains of quartz or, more seldom, feldspar composition, having an isometric rounded shape. The grain size varies from 10 to 50 μm. The surface of grains is usually covered with a thin film of clay particles of micron and sub-micron sizes (Fig. 9.6). Clay particles also often fill the space between silt grains. Sometimes, when the clay fraction is abundant, clay particles may create a non-oriented or slightly oriented matrix, in which silt grains are arranged.

In the course of compaction of such sediments at the stages of early and late diagenesis (to depths of ~ 80 m), the silt grains surrounded by hydrated films of clay particles draw together, with the formation of at first distant and then close coagulation contacts between them. Such clayey-silt sediments have low strength and show well pronounced thixotropic properties, which is characteristic of typical coagulation structures [69].

In the course of catagenesis (at depths > 80 m) under the action of growing geostatic pressure, the effective stresses in the zone of direct contact of silt grains considerably increase. This may result in the plastic deformation of portions of the grain contact area and cause the rise of developed phase contacts and formation of clayey-silt rocks with typical crystallization structure.

Fig. 9.6. Microstructure of a light clay sample N_2^2, depth of recovery 1200 m, the Baku archipelago region; a clay seal of class VI.

A confirmation to this is the high strength of siltstones which usually do not get soaked in water.

A theoretical calculation of the strength of individual contacts in clay rocks composing the seals of classes V–VII was conducted according to the following scheme.

At shallow depths (up to 80 m), when rounded silt grains covered with thin hydrated clay films approach each other at a distance of hundreds to tens of angstrom units, P_1 is mainly determined by the gradually growing molecular attraction.

To calculate the forces of molecular attraction for two spherical particles of the same radius, one can use Hamaker's formula [41]:

$$F = A \times R/12 \times H^2 \tag{9.13}$$

where A = Hamaker's constant depending on the nature of interaction between bodies, R = radius of spheres, and H = clearance between spheres.

The constant A for quartz is 4.5×10^{-20} J [41].

The radius R for spherical particles, i.e. the size of silt grains was determined by the authors from data on the granulometric composition of clay-silt sediments at depths up to 80 m and taken equal to 7.5 μm.

An approximate calculation performed according to Formula (9.13) for clearance $H \geqslant 10$ Å has shown that in clay-silt sediments the strength of close coagulation contacts between silt quartz grains arises from $P_1 \sim 3.3 \times 10^{-9}$ N at shallow depth of 50 m up to $P_1 \sim 2.5 \times 10^{-8}$ N, at depth about 80 m.

To estimate P_1 at depths $h > 80$ m, another scheme was applied, based on the dependence of cohesion forces on the pressing forces between contacting microcrystals. Such a dependence has been experimentally obtained by R.K. Yusupov [111].

The results of investigations by R.K. Yusupov are shown in Figure 9.7. It presents the differential distribution functions of contacts between the microcrystals of talc and vermiculite according to their strength depending on the pressing force varying from 3×10^{-8} to 5×10^{-4} N.

As follows from these data, at pressing forces up to 5×10^{-6} N, the strength of an individual contact remains low ($\sim 10^{-8}$ N), i.e. is mainly determined by molecular forces.

When a certain critical pressing force is reached, which is about 5×10^{-5} N for layered silicates, the number of stronger transitional contacts with $P_1 \approx 10^{-7}$ N and more on the histograms noticeably increases. As the pressing force increases, the portion of such contacts grows, and so does also their average strength, which may be related to the increase in the contact area. At pressing forces up to 5×10^{-4} N, the «weak» contacts practically disappear in talc but are partially preserved in vermiculite. This is explained by the fact that talc is a softer mineral and the number of phase contacts formed in it under a certain loading is greater than in vermiculite.

Thus, at $f_p < 5 \times 10^{-5}$ N, the less strong contacts prevail, which are determined mainly by the action of molecular forces. At $f_p > 10^{-5}$ N, a quick growth of transitional and then also phase contacts of crystalline nature is observed.

To estimate the strength of individual contacts in the seals of V–VII classes at depths exceeding 80 m, the authors have used the data by R.K. Yusupov on vermiculite (Fig. 9.7, *a*), which is the closest analogue of clay mineral particles covering the surface of quartz grains in siltstones.

The calculation of P_1 was performed in the following way. From the known depth of occurrence (h) and density (ρ) of samples representing the clay seals of classes V–VII, the geostatic pressure (P_{gs}) at this depth was determined:

$$P_{gs} = \rho \times g \times h \tag{9.14}$$

The calculations have shown that geostatic pressure in such rocks may vary from 3.25 MPa at a depth of $\sim$200 m to 71.44 MPa at a depth of $\sim$3000 m.

For the same samples, a calculation of the number of contacts in a unit area of the failure section χ has been performed based on the bi-dispersed model. The method of calculation will be described in more detail in the following paragraph.

The pressing force f_p was determined as follows:

$$f_p = \frac{P_{gs}}{\chi} = \frac{\rho \times g \times h}{\chi} \tag{9.15}$$

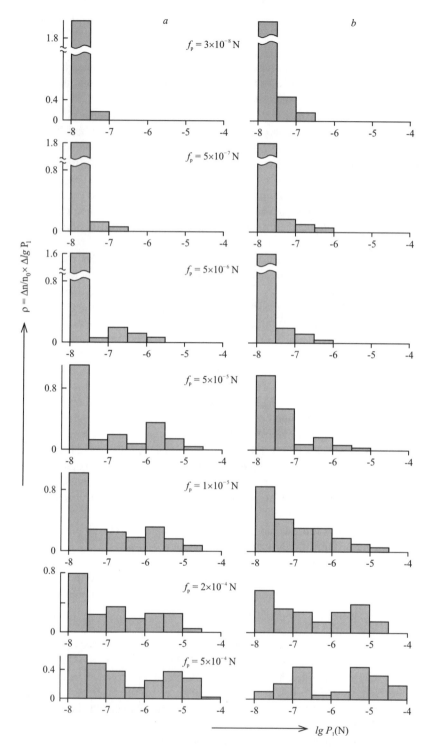

Fig. 9.7. Differential distribution functions of individual contact strength P_1 between the microcrystals of vermiculite (*a*) and talc (*b*) depending on the pressing force f_p ranging from 3×10^{-8} N to 5×10^{-4} N.

This approximate calculation has shown that the pressing force between silt grains covered with clay particles in such a fine grained system grows from 8.23×10^{-6} N at a depth of ~ 200 m to 1.95×10^{-4} N at a depth of ~ 3000 m.

Then, from the histogram (Fig. 9.7, a), an average strength of individual contacts P_1 was estimated, which varied from 3.4×10^{-8} N at a depth of ~ 200 m to 1.2×10^{-6} N at a depth of ~ 3000 m.

The obtained data have allowed the authors to construct a theoretical curve for the change of individual contact strength with depth (Fig. 9.5, curve 3). The obtained relation is described by a quadric polynomial equation of the following form:

$$P_1 = 7.09346 \times 10^{-9} + 1.02161 \times 10^{-10} \times h + 9.92947 \times 10^{-14} \times h^2$$

(9.16)

The high values of the correlation coefficient $R = 0.99$ indicate a very close correlation between the parameters P_1 and h.

An analysis of the relation (9.16) shows that up to the depth of 80 m, P_1 sharply increases to the value of $\sim 1.6 \times 10^{-8}$ N. This evidences that at the stages of early and late diagenesis, transformation of distant coagulation contacts into the close ones occurs.

Further subsidence of a clay-silt sediment is also accompanied by rather intense growth of P_1, which begins to slow down at a depth of ~ 700 m, and at a depth of ~ 900 m P_1 attains the values of $\sim 1.8 \times 10^{-7}$ N. So, at the stage of early catagenesis, the completion of the process of transformation of close coagulation contacts into transitional ones is observed.

At depths up to ~ 2000–2100 m, the growth of P_1 slows down considerably and it attains the values of $\sim 6.7 \times 10^{-7}$ N. This indicates that in the course of middle catagenesis, the transitional contacts between structural elements in the rocks under consideration become gradually transformed into the phase contacts.

When the rocks subside to greater depths, up to ~ 2600 m, a less intense growth of the individual contact strength to the values of $\sim 9.4 \times 10^{-7}$ N is observed. Such a trend of the $P_1 = f(h)$ relation indicates that at high geostatic pressures and elevated temperatures at the stage of late catagenesis, the contact area between structural elements increases significantly and more strong phase contacts of the crystallization type form.

9.2. CALCULATION OF THE CHANGE IN THE NUMBER OF CONTACTS IN CLAY SEALS WITH DEPTH

9.2.1. SEALS OF CLASSES I–II AND III–IV

Taking into account the anisometric shape and the character of change in the degree of orientation of clay particles with depth, the calculation of the number of contacts between structural elements in a unit area of the failure section χ was conducted using the twisting card house model (see Section 8.2, Formula (8.13)).

To make such an estimate, the knowledge of the principal micromorpho-logical parameters of the model is required: the particle length «a» and the angle between contacting particles «θ».

As has been already noted in Section 9.1.1, data on the change of angle θ with depth h have been obtained using a quantitative computer analysis of clay rock microstructure from their SEM images. Data on the average particle size «a» were obtained in the course of statistical processing of the granulometric analysis results and tested using microstructural investigations in SEM. Relations $a = f(h)$ and $\theta = f(h)$ were given earlier in Figures 9.1 and 9.2.

Graphs showing the change in the number of contacts χ with the depth of occurrence h for clay seals of classes I–II and III–IV are given in Figure 9.8 (curves 1, 2).

For these clay seals, the relation $\chi = f(h)$ is described by a power-function equation:

1. For clay seals of classes I–II:

$$\chi = 1.41079 \times 10^{-9} - 4.65052 \times 10^{8} \times h^{0.0635} \tag{9.17}$$

2. For clay seals of classes III–IV:

$$\chi = 8.2273 \times 10^{-8} - 3.27542 \times 10^{7} \times h^{0.26888} \tag{9.18}$$

Both of the equations are characterized by very close correlation between the parameters χ and h, which is evidenced by the high values of correlation coefficients $R = 0.99$.

As is seen from the graph (Fig. 9.8, curves 1, 2), as the depth of occurrence grows, the number of contacts in the seals of classes I–II and III–IV decreases in spite of the porosity reduction, which is caused by the growth of the structural element size with depth.

The most intense decrease of χ with depth h (from $\sim$7.4 $\times$ 10^8 contacts/cm^2 at $h = 30$ m to $\sim$5.2 $\times 10^8$contacts/cm^2 at $h = 3800$ m) occurs in the seals of classes III–IV (Fig. 9.8, curve 2), which is determined by the peculiarities of the mineral composition and by the character of compaction of these clay rocks.

The cause is that the clay seals of classes III–IV are composed of more coarse kaolinite, hydromica, and chlorite particles, whose size grows more quickly with depth than in the seals of classes I–II. Besides, the process of angle θ decrease is observed to slow down at depths over $\sim$2000 m, because the formation of transitional and phase contacts already occurs there to add more rigidity to the clay rock structure.

9.2.2.   SEALS OF CLASSES V–VII

The clay seals of classes V–VII (silty clays and siltstones) are characterized by the presence of large silt grains and small clay particles. The clay particles in these rocks usually form a thin film («clay coats») on the surface of silt grains.

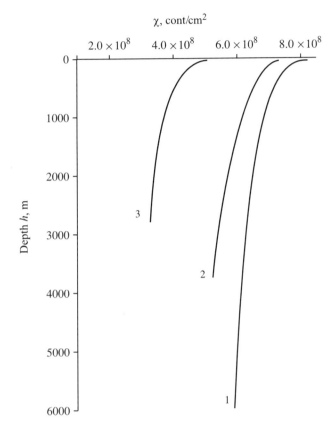

Fig. 9.8. Change in the number of contacts in a unit area of the failure section χ depending on the depth of occurrence h for clay seals: 1 – of classes I–II; 2 – of classes III–IV; 3 – of classes V–VII.

All this allows one to use the bi-dispersed model (see Section 8.2) for the calculation of the number of contacts in such fine grained rocks. So, the calculation of χ in the clay seals of classes V–VII was performed using Formula (8.10).

For the calculation, the following parameters of the model should be known: porosity (n), coordination number (z), average radii of large (R) and small (r) particles, average density of minerals composing the large (ρ_R) and small (ρ_r) particles, average weight contents of these particles $(\varphi_R$ and $\varphi_r)$. Data on the porosity change with depth were taken from published works and given earlier in Figure 7.9. The coordination number was calculated based on the rock porosity data according to Formula (8.5).

A theoretical calculation of parameter χ was accomplished by the authors for quartz silt grains with $\rho_R = 2.65\,\text{g/cm}^3$ covered with clay particles with $\rho_r = 2.70\,\text{g/cm}^3$ [3].

The sizes of particles of the silt and clay fractions and their change with depth were determined using statistical processing of available from literature data on the granulometric analysis of clay rock samples representing the seals of classes V–VII [255, 256].

The obtained results are given in Figure 9.9. It follows from them that the size of particles of both large (curve 1) and small (curve 2) fractions increases with depth.

The average weight contents of large and small particles were also estimated from the granulometric composition data. On the whole, an increase in the silt fraction content with depth was observed. In the calculations for depths up to 1000 m, φ_R and φ_r were taken in equal quantities, 50% each, while for greater depths, $\varphi_R = 60\%$ and $\varphi_r = 40\%$.

The results of calculation of the contact number in clay seals of classes V–VII and their change with depth are presented in Figure 9.8 (curve 3). Here, the relation $\chi = f(h)$, as well as for the seals of classes I–II and III–IV, is described by a power-function equation:

$$\chi = 4.04755 \times 10^9 - 3.39227 \times 10^9 \times h^{0.01572} \tag{9.19}$$

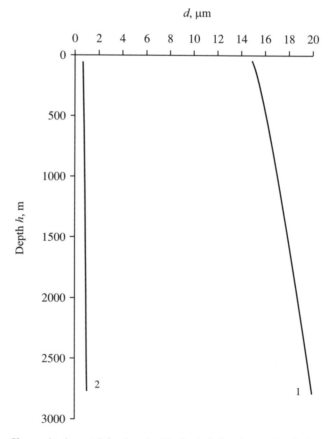

Fig. 9.9. Change in the particle size d with depth h for clay seals of classes V–VII: 1 – silt particles; 2 – clay particles.

The high value of correlation coefficient (R = 0.97) indicates the close inter-relation between the parameters χ and h.

9.3. THEORETICAL CALCULATION OF TENSILE AND COMPRESSIVE STRENGTH OF CLAY SEALS

The possibility of theoretical estimation of the individual contact strength P_1 and the number of contacts χ and the knowledge of their change depending on depth h allow one to use the principal equation in the physicochemical mechanics of dispersed media (8.1) in order to estimate the tensile strength of clay rocks (P_t) and their compressing strength (P_c). Besides, the tensile strength is not just a strength characteristic of a rock, but it may also be used for estimating such an important petrophysical parameter as the formation breakdown pressure.

9.3.1. THEORETICAL CALCULATION OF THE TENSILE STRENGTH OF CLAY SEALS

The results of theoretical calculation of tensile strength for clay seals of classes I–II, III–IV, and V–VII are given in Figure 9.10. It should be noted that the theoretical relations $P_t = f(h)$ for all of the seal classes are described by quadric polynomial equations:

1. For clay seals of classes I–II:

$$P_t = 0.0997 + 3.3142 \times 10^{-4} \times h + 1.0144 \times 10^{-7} \times h^2; \quad R = 0.99$$
$$(9.20)$$

2. For clay seals of classes III–IV:

$$P_t = 0.06267 + 6.67745 \times 10^{-4} \times h + 2.32493 \times 10^{-8} \times h^2; \quad R = 0.99$$
$$(9.21)$$

3. For clay seals of classes V–VII:

$$P_t = 0.02419 + 4.9429 \times 10^{-4} \times h + 9.14635 \times 10^{-8} \times h^2; \quad R = 0.99$$
$$(9.22)$$

When using the obtained theoretical relations $P_t = f(h)$ for estimating the tensile strength of real clay rocks, it should be taken into account that under natural conditions, the highest screening properties are preserved in clay seals of classes I–II in the interval from ~400–500 m to the maximal depths of ~6000 m; in seals of classes III–IV – from ~1000 m to ~2500 m, and in seals of classes V–VII–from ~1200 m to ~1800 m.

Below the indicated depths, at which clay seals lose their screening properties owing to the development of microfractures in them, the tensile strength of clay rocks under real conditions may be lower than the calculated values.

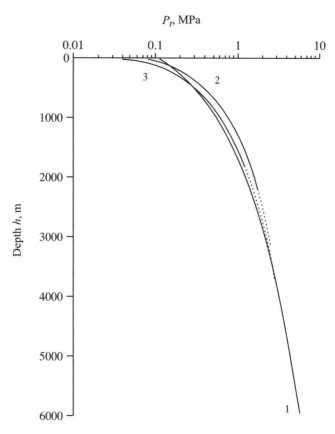

Fig. 9.10. Theoretical curves of change in the tensile strength P_t with depth h for clay seals: 1 – of classes I–II; 2 – of classes III–IV; 3 – of classes V–VII.

In connection with these restrictions, the authors believe that, based on the obtained theoretical relations $P_t = f(h)$, the prediction of tensile strength for the clay seals of classes I–II is possible throughout the whole depth interval up to 5000–6000 m, while for the clay seals of classes III–IV and V–VI it may be done only to the depths of 2500 and 1800 m, respectively. These segments are dotted in Figure 9.10 (curves 2, 3).

A comparative analysis of curves $P_t = f(h)$ (Fig. 9.10) shows that in the depth interval from ~200 m to ~2500 m, the clay seals of classes III–IV have the highest tensile strength. The explanation for this is that at these depths, such clay sediments have already passed the stage of early catagenesis and complete the stage of middle catagenesis, with transitional contacts being dominant in them and phase contacts beginning to arise.

The clay seals of classes I–II at these depths have reached in their geological evolution only the stage of middle catagenesis, so that the close coagulation and transitional contacts may still be present in them determining the lower tensile strength of such deposits. However, the advantage of the I–II class seals is that they ensure the high screening properties within a very large interval of depths, practically up to 6000 m. At the same time, the tensile strength of these seals gradually increases with depth, while clay rocks forming the seals of classes III–IV and V–VII lose their isolation properties already at the depths of ~2500 and ~1800, respectively.

The tensile strength of the V–VII class clay seals at depths of 500–1800 m has an intermediate value between those for the seals of classes III–IV and I–II, which is also determined by the formation in them within this depth interval of strong transitional and phase contacts. However, due to the large size of structural elements in these rocks, the number of contacts in them turns to be less than in the seals of classes III–IV, and so their tensile strength turns to be lower.

9.3.2. THEORETICAL CALCULATION OF THE COMPRESSIVE STRENGTH OF CLAY SEALS

The tensile strength of clay rocks is connected with their compressing strength [3, 22, 93]. In rock mechanics, there is such an index as rock brittleness (fragility) K_f determined by the ratio of the uniaxial compression strength to the tensile strength [22]:

$$K_f = P_c/P_t \qquad (9.23)$$

An analysis of published data [3, 93] and experimental investigations by the authors [51, 54, 69] show that the ratio P_c/P_t may vary considerably in clay rocks with different types of structure. Thus, $P_c/P_t \approx 5$ is characteristic of young clay sediments and clays having a weak degree of lithification with coagulation structures, in which distant and close coagulation contacts prevail. For clay rocks of a medium lithification degree with transitional structures characterized by the presence of mainly transitional contacts, $P_c/P_t \approx 8$. And, at last, for clay rocks of a high lithification degree – argillites, siltstones, and clay schists with the crystallization-cementation structures, which are characterized by the presence of phase contacts, $P_c/P_t \approx 10$.

These relations between P_c and P_t were used by the authors for theoretical estimation of the uniaxial compression strength of clay seals from the calculated data on their tensile strength.

Figure 9.11 shows theoretical relationships between the uniaxial compression strength of clay seals and the depth of rock occurrence. It follows from them, that the $P_c = f(h)$ dependence for all of the seal classes is also described by a quadric polynomial equation:

1. For clay seals of classes I–II:

$$P_c = 0.4438 + 0.00328 \times h + 1.04629 \times 10^{-6} \times h^2; \quad R = 0.99 \qquad (9.24)$$

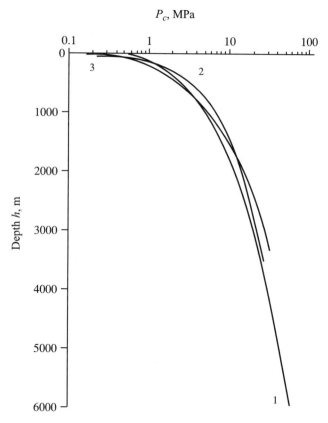

Fig. 9.11. Theoretical curves of change in the compression strength P_c with depth h for clay seals: 1 – of classes I–II; 2 – of classes III–IV; 3 – of classes V–VII.

2. For clay seals of classes III–IV:

$$P_c = 0.13252 + 0.00596 \times h + 4.84865 \times 10^{-7} \times h^2; \quad R = 0.99 \quad (9.25)$$

3. For clay seals of classes V–VII:

$$P_c = 0.06575 + 0.00331 \times h + 1.917 \times 10^{-6} \times h^2; \quad R = 0.99 \quad (9.26)$$

In spite of the fact that compression strength is less sensitive to the processes of microfracturing growth in clay rocks at great depths, it should be suggested that the correct prediction of P_c for real clay seals of classes III–IV and V–VII may only be done to the depths at which their screening properties begin to lower, i.e. to 2500 m and 1800 m, respectively. Therefore, these portions on the graphs (Fig. 9.11, curves 2, 3) are dotted.

One of the important structural parameters of clay rocks, which is closely connected with their depth of occurrence, is porosity. Therefore, it is of practical interest to find a theoretical dependence of the change in the compression

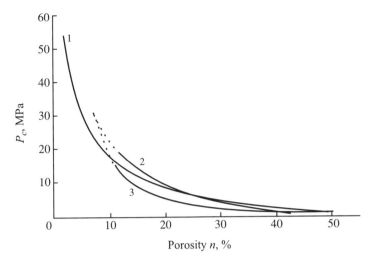

Fig. 9.12. Theoretical curves of change in the compressive strength P_c depending on porosity n for clay seals: 1 – of classes I–II; 2 – of classes III–IV; 3 – of classes V–VII.

strength P_c on the porosity of clay seals (Fig. 9.12). This dependence is described for all of the seal classes by a power-function equation:

1. For clay seals of classes I–II:

$$P_c = -14.8832 + 97.74111 \times n^{-0.47104}; \quad R = 0.99 \qquad (9.27)$$

2. For clay seals of classes III–IV:

$$P_c = -18.19582 + 144.41892 \times n^{-0.55309}; \quad R = 0.99 \qquad (9.28)$$

3. For clay seals of classes V–VII:

$$P_c = -0.81453 + 1309.90996 \times n^{-1.83716}; \quad R = 0.99 \qquad (9.29)$$

The character of change in the relations $P_c = f(n)$ is in accordance with the trends of porosity (n) and strength (P_c) change with depth h.

Dotted lines on curves $P_c = f(n)$ for the seals of classes III–IV and V–VII (Fig. 9.12, curves 2, 3) show the trend of these relations for the porosity values corresponding to the depths of 2500 m and 1800 m, as theoretical data on P_c at this segments may be too high and not corresponding to the compression strength of real samples because of microfracturing processes developed in them.

9.3.3. «SEAL TEST» PROGRAM FOR ESTIMATING THE MECHANICAL STRENGTH OF CLAY SEALS

The obtained theoretical dependence of the strength of an individual contact between particles on the occurrence depth of different-class clay seals ($P_1 = f(h)$)

has formed a basis of the specialized computer «Seal Test» program for estimating the mechanical strength of clay seals.

The program works in Microsoft Windows 95/98/NT operating system. To
begin working with the program, it is necessary to copy the «Stest.exe» file
from the applied disk into any directory of the PC hard disk for further running.

The main window of the program consists of three parts:

– in the upper part of the main window there is a tab control aiming to choose
  the seal classes (Figs. 9.13(6)–9.15(6)), controls for the input and editing of
  initial parameters of the calculation model;
– lower, there are windows displaying the results of tensile strength P_t
  (Figs. 9.13(2)–9.15(2)) and compressive strength P_c (Figs. 9.13(4)–9.15(4))
  calculations in accordance with the active class of clay seals and the input
  initial parameters;
– in the lower part of the main window, the program control buttons are
  located: <Calculate> – the button for beginning the calculations for the
  chosen class of clay seals (Figs. 9.13(1)–9.15(1)); <Clear> – the button for

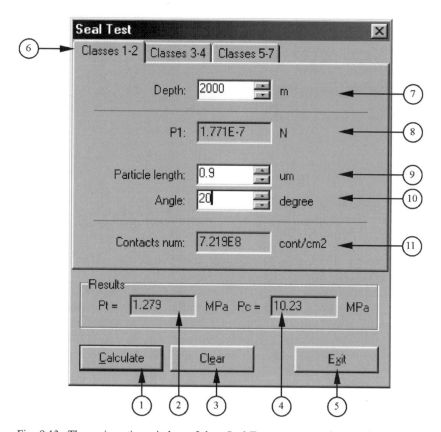

Fig. 9.13. The main active window of the «Seal Test» program when performing the
calculations of tensile and compressive strength for clay seals of classes I–II.

clearing the results of previous calculations and initializing the initial parameters of the active class of seals (Figs. 9.13(3)–9.15(3)); <Exit> – the button for closing the program (Figs. 9.13(5)–9.15(5)).

The work with the program begins with the choice of the clay seal class. To do this, the user must click one of the three tabs in the upper part of the main window of the program (Figs. 9.13(6)–9.15(6)). The first tab corresponds to clay seals of classes I and II, the second tab – to clay seals of classes III and IV, and the third tab – to classes V–VII.

Then the user must input the initial parameters of the calculation model. For different types of seals, different initial parameters are set.

For clay seals of classes I–II and III–IV, the following parameters must be input:

– depth of clay seal occurrence, in meters (Figs. 9.13(7)–9.14(7));
– average length of clay particles, in microns (Figs. 9.13(9)–9.14(9));
– angle of inclination between contacting particles, in degrees (Figs. 9.13(10)–9.14(10)).

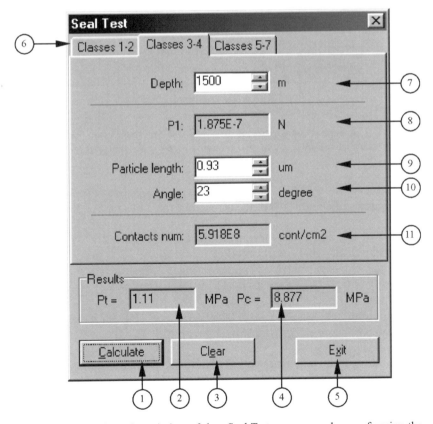

Fig. 9.14. The main active window of the «Seal Test» program when performing the calculations of tensile and compressionve strength for clay seals of classes III–IV.

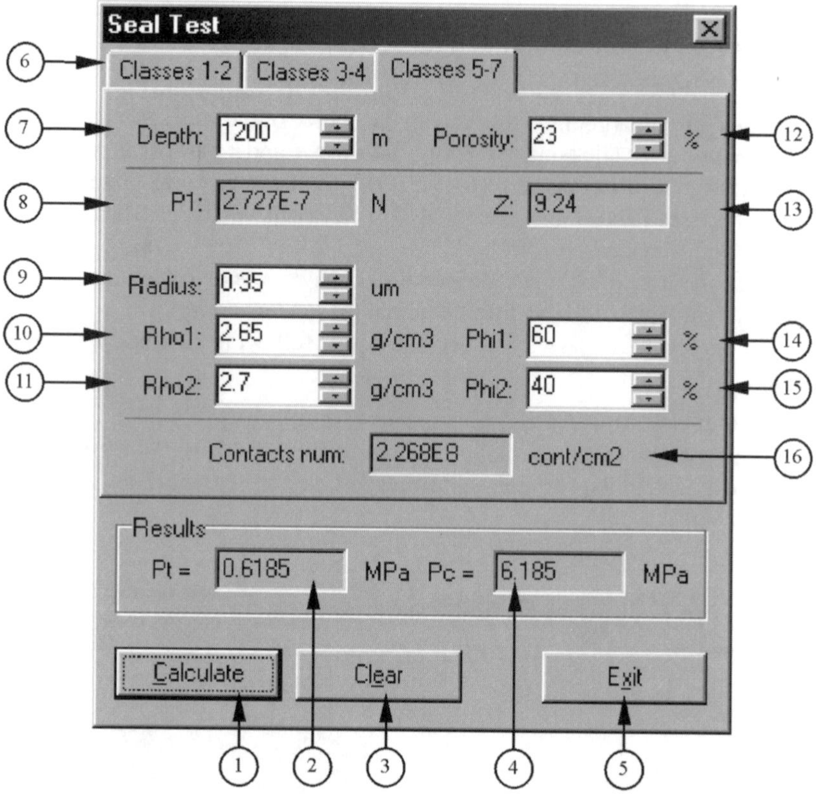

Fig. 9.15. The main active window of the «Seal Test» program when performing the calculations of tensile and compressive strength for clay seals of classes V–VII.

For clay seals of classes V–VII, the parameters to be input are as follows:

– depth of clay seal occurrence, in meters (Fig. 9.15(7));
– porosity of clay seal, in per cent (Fig. 9.15(12));
– average radius of small (clay) particles (Fig. 9.15(9));
– density of large (silt) particles ρ_1 (Fig. 9.15(10)) and density of small (clay) particles ρ_2 (Fig. 9.15(11)), in g/cm^3;
– concentration by weight of large (silt) particles φ_1 (Fig. 9.15(14)) and small (clay) particles φ_2 (Fig. 9.15(15)), in per cent.

On the basis of the input parameters, after the <Calculate>button is pressed (Figs. 9.13(1)–9.15(1)), the calculation of parameters in question is performed:

– P_t – tensile strength, MPa (Figs. 9.13(2)–9.15(2));
– P_c – compressive strength, MPa (Figs. 9.13(4)–9.15(4)).

Besides the principal parameters in question, a number of intervening parameters are determined, which are required for calculations in accordance with the calculation model:

– P_1 – strength of an individual contact, in newtons (Figs. 9.13(8)–9.15(8));
– the number of contacts in a unit area of the failure section, in contacts/cm^2 (Figs. 9.13(11)–9.14(11); Fig. 9.15(16));
– coordination number z (Fig. 9.15(13)).

These parameters are displayed on the screen with the purpose of information.

To perform the repeated calculations, the user must repeat the above-described consecutive actions. Before doing this, the initial data and the results of previous calculations may be cleared by pressing the <Clear>button (Figs. 9.13(3)–9.15(3)).

The work with the program is finished by pressing the <Exit>button (Figs. 9.13(5)–9.15(5)).

10

Comparison of Theoretical Calculation Results and Experimental Data on the Mechanical Strength of Clay Seals

10.1. EXPERIMENTAL DATA BASE

To verify the results of theoretical calculation of the strength characteristics of clay seals, the authors have collected samples of clay rocks having different age, genesis, and degree of lithification and recovered from various regions of the former USSR. This collection includes 21 samples of clay seals ranking with classes I–II, 43 samples of clay seals ranking with classes III–IV, and 8 samples of seals ranking with classes V–VII. Unfortunately, scarce information is available in the literature on the composition and properties of V–VII class clay seals, so the authors had to do with such a little sampling. There were the following data for all of the samples: place of recovery, age, depth of occurrence h, results of granulometric and mineral analyses, total porosity n, and uniaxial compressive strength P_c. All the above data on the samples have been reduced to a data base presented in Tables 10.1–10.3.

10.2. COMPARISON OF THEORETICAL AND EXPERIMENTAL DATA

To estimate the trustworthiness of the theoretical calculations of mechanical strength for the clay seals of different classes, a test of the results obtained has been carried out.

Graphs presented in Figures 10.1–10.3 show the compressive strength dependence on the depth of occurrence for real samples of clay seals of classes I–II, III–IV, and V–VII, respectively. Unfortunately, as the sampling of clay seals of classes I–II and V–VII was rather small, the authors could not have constructed experimental curves $P_c = f(h)$ for the entire depth interval, wherein these seals had good isolation properties. So, the experimental data were restricted for the seals of classes I–II by the depth of ~ 2700 m and for the seals of classes V–VII by the depth of ~ 1200 m.

Table 10.1. Experimental data base on the samples of clay seals of classes I–II.

No.	The place of sample recovery	Age	Rock type	Depth of occurrence, h, m	Clay seal class	Total porosity, n, %	Uniaxial compressive strength, P_c, MPa	Silt/clay fraction ratio	Swelling/ non-swelling mineral ratio $M + ML / I + CHL + K*$	Reference
1	2	3	4	5	6	7	8	9	10	11
1	Moscow syneclise, near the Moscow coal basin	C_1	Clay	120	II	36	1.23	0.3	2.0	[257]
2	South-Caspian oil- and gas-bearing region, Baku archipelago	Q_1^{t-b}	Clay	145	II	41.9	1.27	0.3	2.0	[255]
3	South-Caspian oil- and gas-bearing region, Baku archipelago	Q_1^{t-b}	Clay	151	II	41	1.18	0.3	2.0	[255]
4	South-Caspian oil- and gas-bearing region, Baku archipelago	N_2^{3ap}	Clay	277	II	42.5	1.86	0.2	2.0	[255]
5	South-Caspian oil- and gas-bearing region, Baku archipelago	N_2^{3ap}	Clay	281	II	38.7	1.81	0.3	2.0	[255]
6	South-Caspian oil- and gas-bearing region, Baku archipelago	N_2^{3ap}	Clay	283	II	35.5	1.93	0.3	2.0	[255]
7	South-Caspian oil- and gas-bearing region, Baku archipelago	N_2^{3ap}	Clay	295	II	42	0.89	0.2	2.0	[255]
8	South-Caspian oil- and gas-bearing region, Baku archipelago	N_2^2	Clay	615	II	31.8	3.5	0.3	1.5	[255]
9	South-Caspian oil- and gas-bearing region, Baku archipelago	N_2^2	Clay	724	II	33.6	3.42	0.3	1.5	[255]

10	South-Caspian oil- and gas-bearing region, Baku archipelago	N_2^2	Clay	745	II	33	3.05	0.3	1.5	[255]
11	West Siberia, Barabin 1-R stratigraphic test	K_2t	Clay	750	II	11.7	19	0.3	3.0	[256]
12	West Siberia, Tim 1-R stratigraphic test	K_2t	Clay	790	II	14	2.46	0.3	4.0	[256]
13	South-Caspian oil- and gas-bearing region, Baku archipelago	N_2^2	Clay	810	II	31.3	2.75	0.2	1.5	[255]
14	West Siberia, Novo-Vasyugansk well	K_2	Clay	880	II	11.5	3.85	0.1	3.0	[256]
15	South-Caspian oil- and gas-bearing region, Baku archipelago	N_2^2	Clay	892	II	27.8	4.2	0.3	1.5	[255]
16	South-Caspian oil- and gas-bearing region, Baku archipelago	N_2^2	Clay	933	II	29.4	3.5	0.3	1.5	[255]
17	South-Caspian oil- and gas-bearing region, Baku archipelago	N_2^2	Clay	974	II	27.2	3.5	0.3	1.5	[255]
18	South-Caspian oil- and gas-bearing region, Baku archipelago	N_2^2	Clay	1108	II	27.9	3.4	0.3	1.5	[255]
19	Okhotsk oil- and gas-bearing province	$Pg-N$ N	Clay	2000	II	19	12	0.2	2.5	[109]
20	Pre-Caucasus region, Krasnodar territory	K_1	Argillite	2020	II	22	13	0.3	1.5	[258]
21	Pre-Caucasus region, West-Kubanian district	$Pg_3 - N_1^1$	Clay	2450	II	22	11	0.1	3.0	[92]

*Note: M – montmorillonite, ML – mixed-layer minerals, I – illite, CHL – chlorite, K – kaolinite.

(Table 10.1. continued)

Table 10.2. Experimental data base on the samples of clay seals of classes III–IV

No.	The place of sample recovery	Age	Rock type	Depth of occurrence, h, m	Clay seal class	Total porosity, n, %	Uniaxial compressive strength, P_c, MPa	Silt/clay fraction ratio	Swelling/non-swelling mineral ratio $M + ML / I + CHL + K*$	Reference
1	South-Caspian oil- and gas-bearing region, Baku archipelago	Q_{2-4}	Clay	25	IV	42.1	0.16*	0.7	0.8	[255]
2	Volga-Ural oil- and gas-bearing region, environs of Ul'yanovsk	$K_1 b$	Clay	30	IV	45	0.95	0.4	0.7	[257]
3	Moscow syneclise, near the Moscow coal basin	C_1	Clay	30	III	46	1.05	0.6	0.7	[257]
4	Volga-Ural oil- and gas-bearing region, environs of Ul'yanovsk	K_1	Clay	40	IV	49	0.75	0.7	0.7	[257]
5	Moscow syneclise, near the Moscow coal basin	C_1	Clay	50	III	31	0.75	0.7	0.6	[257]
6	South-Caspian oil- and gas-bearing region, Baku archipelago	Q_{2-4}	Clay	55	IV	42.2	0.2	0.7	0.8	[255]
7	Volga-Ural oil- and gas-bearing region, near Goloshubikha village	$P_2 t$	Clay	60	IV	48.8	0.6	0.5	0.7	[257]
8	South-Caspian oil- and gas-bearing region, Baku archipelago	Q_{2-4}	Clay	65	IV	43.6	0.17	0.6	0.9	[255]
9	South-Caspian oil- and gas-bearing region, Baku archipelago	Q_{2-4}	Clay	71	IV	44.8	0.54	0.6	0.9	[255]
10	Volga-Ural oil- and gas-bearing region, near Goloshubikha village	$P_2 t$	Clay	80	IV	47.1	0.8	0.5	0.7	[257]
11	Volga-Ural oil- and gas-bearing region, near Goloshubikha village	$P_2 t$	Clay	100	IV	43.5	0.8	0.5	0.7	[257]

No.	Location									
12	Conjunction of the Voronezh arch and the Dnieper-Donets basin, Belgorod district, Gostintsevo settlement	J_{2-3}	Clay	100	IV	51	0.3	0.5	0.7	[123]
13	South-Caspian oil- and gas-bearing region, Baku archipelago	Q_1^{t-b}	Clay	109	III	41.2	0.48	0.8	1.0	[255]
14	South-Caspian oil- and gas-bearing region, Baku archipelago	Q_1^{t-b}	Clay	127	III	38.4	0.5	0.6	1.0	[255]
15	Volga-Ural oil- and gas-bearing region, environs of Vasil'sursk	P_2t	Clay	150	IV	39	0.9	0.4	0.7	[257]
16	Conjunction of the Voronezh arch and the Dnieper-Donets basin, Belgorod district, Gostintsevo settlement	J_{2-3}	Clay	175	IV	49	1	0.7	0.7	[123]
17	South-Caspian oil- and gas-bearing region, Baku archipelago	N_2^{3ap}	Clay	187	IV	39.9	1.27	0.7	1.0	[255]
18	Conjunction of the Voronezh arch and the Dnieper-Donets basin, Belgorod district, Gostintsevo settlement	J_{2-3}	Clay	210	III	42	1.8	0.5	0.7	[123]
19	South-Caspian oil- and gas-bearing region, Baku archipelago	N_2^{3ap}	Clay	253	III	43.3	0.75	0.4	1.0	[255]
20	South-Caspian oil- and gas-bearing region, Baku archipelago	N_2^{3ap}	Clay	301	III	39	2.2	0.5	1.0	[255]
21	South-Caspian oil- and gas-bearing region, Baku archipelago	N_2^{3ap}	Clay	313	IV	39.8	1.91	0.4	1.0	[255]

(Table 10.2. continued)

Table 10.2. Continued.

No.	The place of sample recovery	Age	Rock type	Depth of occurrence, h, m	Clay seal class	Total porosity, n, %	Uniaxial compressive strength, P_c, MPa	Silt/clay fraction ratio	Swelling/ non-swelling mineral ratio $M + ML / I + CHL + K*$	Reference
1	2	3	4	5	6	7	8	9	10	11
22	South-Caspian oil- and gas-bearing region, Baku archipelago	N_2^{3ap}	Clay	343	IV	35.5	3.57	0.6	1.0	[255]
23	Conjunction of the Voronezh arch and the Dnieper-Donets basin, Belgorod district, Gostintsevo settlement	J_3km-ox	Argillite	350	III	44	3.9	0.8	0.8	[123]
24	South-Caspian oil- and gas-bearing region, Baku archipelago	N_2^{3ap}	Clay	367	IV	37.1	1.75	0.6	0.8	[255]
25	South-East of the West-Siberian plate, Kuzbass, Prokop'evsk	$C_3 - P_1$	Argillite	410	IV	22.4	4.15	1.0	0.7	[259]
26	South-East of the West-Siberian plate, Kuzbass, Prokop'evsk	$C_3 - P_1$	Argillite	650	III	24.2	5.96	0.7	0.6	[259]
27	South-Caspian oil- and gas-bearing region, Baku archipelago	N_2^2	Clay	665	IV	32.4	3.5	0.6	1.0	[255]
28	South-Caspian oil- and gas-bearing region, Baku archipelago	N_2^2	Clay	759	IV	28.4	3.69	0.7	0.8	[255]
29	West Siberia, Lyushin area, 1-R stratigraphic test	$K2t$	Clay	780	III	26	6	1.0	0.7	[256]
30	South-Caspian oil- and gas-bearing region, Baku archipelago	N_2^2	Clay	843	III	24.6	4.04	1.1	1.0	[255]

No.	Region	Age	Rock	Depth						Ref.
31	South-Caspian oil- and gas-bearing region, Baku archipelago	N_2^2	Clay	899	III	27.6	5.5	0.7	0.8	[255]
32	South-Caspian oil- and gas-bearing region, Baku archipelago	N_2^2	Clay	919	IV	28.1	3.8	0.6	0.9	[255]
33	South-Caspian oil- and gas-bearing region, Baku archipelago	N_2^2	Clay	1018	III	27.5	4.8	0.5	1.0	[255]
34	South-Caspian oil- and gas-bearing region, Baku archipelago	N_2^2	Clay	1050	IV	22.8	3.64	0.8	1.0	[255]
35	South-Caspian oil- and gas-bearing region, Baku archipelago	N_2^2	Clay	1137	III	21	5.38	1.0	0.6	[255]
36	South-Caspian oil- and gas-bearing region, Baku archipelago	N_2^2	Clay	1151	IV	27.4	3.6	1.0	0.7	[255]
37	Conjunction of the Dnieper-Donets basin and Donbass	C_1	Argillite	1200	IV	16.7	8.3	0.8	0.7	[260]
38	Pre-Caucasus region, West-Kubanian foredeep	$Pg_3 - N_2^2$	Clay	1500	III	26	10	0.4	1.0	[92]
39	West Siberia, Zav'yalovo 3-R well	$K_1\,al - ap$	Argillite	1800	III	19.7	11.3	0.4	0.7	[256]
40	Pre-Caucasus region, West-Kubanian foredeep	$Pg_3 -$	Clay	2050	III	17	18.5	0.5	0.7	[92]
41	Pre-Caucasus region, Krasnodar territory	N_2^2 K_1	Argillite	2130	III	20	13	3.0	0.6	[258]
42	Okhotsk oil- and gas-bearing province, northern Sakhalin	$Pg - N$	Argillite	3000	III	14	16.3	0.7	0.8	[109]
43	Okhotsk oil- and gas-bearing province, northern Sakhalin	$Pg - N$	Argillite	4000	III	10	25	1.0	0.7	[109]

Table 10.3. Experimental data base on the samples of clay seals of classes V–VII.

No.	The place of sample recovery	Age	Rock type	Depth of occurrence, h, m	Clay seal class	Total porosity, n, %	Uniaxial compressive strength, P_c, MPa	Silt/clay fraction ratio	Swelling/non-swelling mineral ratio $M + ML/I + CHL + K*$	Reference
1	South-Caspian oil- and gas-bearing region, Baku archipelago	N_2^{3ap}	Clay	451	V	36.0	1.79	0.6	0.5	[255]
2	South-Caspian oil- and gas-bearing region, Baku archipelago	N_2^{3ap}	Clay	463	V	34.0	2.32	0.7	0.5	[255]
3	South-Caspian oil- and gas-bearing region, Baku archipelago	N_2^{3ap}	Clay	499	V	32.5	2.2	0.8	0.6	[255]
4	South-Caspian oil- and gas-bearing region, Baku archipelago	N_2^{3ap}	Clay	535	VI	31.7	2.3	0.9	0.4	[255]
5	South-Caspian oil- and gas-bearing region, Baku archipelago	N_2^{3ap}	Clay	544	VI	30.0	2.51	0.8	0.4	[255]
6	South-Caspian oil- and gas-bearing region, Baku archipelago	N_2^2	Clay	675	V	27.1	2.8	1.1	0.5	[255]
7	South-Caspian oil- and gas-bearing region, Baku archipelago	N_2^2	Clay	773	V	26.0	3.6	0.9	0.5	[255]
8	South-Caspian oil- and gas-bearing region, Baku archipelago	N_2^2	Clay	1200	VI	23.0	6.1	2.0	0.3	[255]

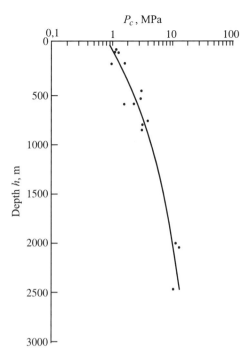

Fig. 10.1. An experimental curve showing the change of compression strength P_c with depth of occurrence h for the samples of I–II class clay seals.

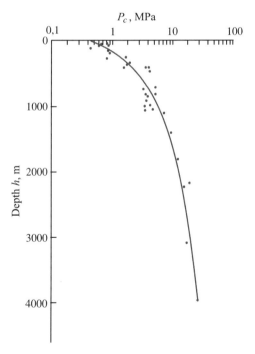

Fig. 10.2. An experimental curve showing the change of compression strength P_c with depth of occurrence h for the samples of III–IV class clay seals.

It follows from the results obtained that all the dependence curves $P_c = f(h)$ constructed on the basis of experimental data (see data base, Tables 10.1–10.3) are described by regression quadric polynomial equations:

1. For clay seals of classes I–II:

$$P_c = 0.69973 + 0.00246 \times h + 1.13662 \times 10^{-6} \times h^2; \quad R = 0.90$$

(10.1)

2. For clay seals of classes III–IV:

$$P_c = 0.19968 + 0.00571 \times h + 4.24267 \times 10^{-7} \times h^2; \quad R = 0.93$$

(10.2)

3. For clay seals of classes V–VII:

$$P_c = 0.71908 + 0.00191 \times h + 2.14669 \times 10^{-6} \times h^2; \quad R = 0.98$$

(10.3)

In performing the computer calculation of the tensile strength P_t and uniaxial compressive strength P_c, one had to know the micromorphological parameters of the models used for estimating the number of contacts χ.

As has been noted earlier, to calculate χ in clay seals of classes I–II and III–IV, data were required on the length of clay particles a and the angle of inclination θ between contacting particles. The parameter a was determined

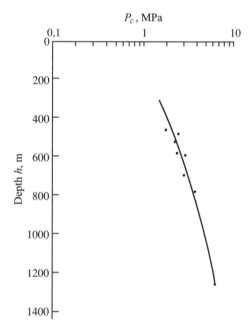

Fig. 10.3. An experimental curve showing the change of compression strength P_c with depth of occurrence h for the samples of V–VII class clay seals.

from the results of granulometric analysis, and the angle θ was estimated using a quantitative microstructural analysis of these samples from SEM images. It should be noted that when samples were not available, the angle θ was determined from their SEM photographs obtained earlier, which were input into the computer with the aid of a scanner and then calculations were performed using the «STIMAN» program. The model parameters a and θ thus obtained and their change with depth are given in Tables 10.4 and 10.5.

In clay seals of classes V–VII, the calculation of the number of contacts was performed based on the bi-dispersed model. In this case, the following model parameters had to be known: porosity (n), average radius of large (R) and small (r) particles, their densities $(\rho_R$ and $\rho_r)$ and weight contents $(\varphi_R$ and $\varphi_r)$. All these values were taken by the authors from published data [255]. They are summarized in Table 10.6.

Figures 10.4–10.6 show the results of calculations of tensile strength P_t and its change with depth accomplished using the «Seal Test» program. As it follows from these calculations, all the obtained regression relations $P_t = f(h)$ are described by quadric polynomial equations:

For clay seals of classes I–II:

$$P_t = 0.08965 + 4.28289 \times 10^{-4} \times h + 6.79919 \times 10^{-8} \times h^2; \quad R = 0.99$$

$$(10.4)$$

Table 10.4. Micromorphological parameters of the «twisting card house» model determined for the clay seals of classes I–II.

No.	Depth of sample occurrence h, m	Length of clay particles a, μm	Angle of inclination between contacting clay particles θ, degrees	Clay seal class
1	120	0.55	60	II
2	145	0.55	57	II
3	151	0.55	56	II
4	277	0.58	49	II
5	281	0.58	49	II
6	283	0.58	49	II
7	295	0.6	48	II
8	615	0.65	36	II
9	724	0.68	34	II
10	745	0.68	33	II
11	750	0.68	32	II
12	790	0.7	32	II
13	810	0.7	32	II
14	880	0.71	31	II
15	892	0.72	30	II
16	933	0.73	30	II
17	974	0.75	29	II
18	1108	0.77	24	II
19	2000	0.90	20	II
20	2020	0.93	20	II
21	2450	1.03	17	II

Table 10.5. Micromorphological parameters of the «twisting card house» model determined for the clay seals of classes III–IV.

No.	Depth of sample occurrence h, m	Length of clay particles a, μm	Angle of inclination between contacting clay particles θ, degrees	Clay seal class
1	25	0.52	82	IV
2	30	0.52	80	IV
3	30	0.52	78	III
4	40	0.52	75	IV
5	50	0.53	75	III
6	55	0.53	73	IV
7	60	0.54	70	IV
8	65	0.54	67	IV
9	71	0.54	67	IV
10	80	0.55	65	IV
11	100	0.56	61	IV
12	100	0.54	62	IV
13	109	0.55	60	III
14	127	0.57	56	III
15	150	0.58	55	IV
16	175	0.58	53	IV
17	187	0.6	49	IV
18	210	0.64	45	III
19	253	0.65	44	III
20	301	0.66	43	III
21	313	0.66	43	IV
22	343	0.68	41	IV
23	350	0.68	41	III
24	367	0.68	40	IV
25	410	0.7	39	IV
26	650	0.76	32	III
27	665	0.76	30	IV
28	759	0.8	29	IV
29	780	0.8	29	III
30	843	0.8	27	III
31	899	0.83	28	III
32	919	0.83	28	IV
33	1018	0.85	27	III
34	1050	0.86	27	IV
35	1137	0.85	26	III
36	1151	0.85	26	IV
37	1200	0.89	25	IV
38	1500	0.93	23	III
39	1800	0.98	21	III
40	2050	1.01	20	III
41	2130	1.05	19	III
42	3000	1.19	15	III
43	4000	1.35	11	III

Table 10.6. Parameters of the bi-dispersed model determined for the clay seals of classes V–VII.

No.	Depth of sample occurrence h, m	Total porosity n, %	Radius of silt grains R, μm	Radius of clay particles r, μm	Weight content of silt grains φ_R,%	Weight content of clay particles φ_r,%	Density of silt grains ρ_R, g/cm³	Density of clay particles ρ_r, g/cm³	Clay seal class
1	451	36	8	0.3	50	50	2.65	2.70	V
2	463	34	8.5	0.3	50	50	2.65	2.70	V
3	499	32.5	8.5	0.33	50	50	2.65	2.70	V
4	535	31.7	8.5	0.33	50	50	2.65	2.70	VI
5	544	30	8.5	0.33	50	50	2.65	2.70	VI
6	675	27.1	8.5	0.35	50	50	2.65	2.70	V
7	773	26	8.5	0.35	50	50	2.65	2.70	V
8	1200	23	8.5	0.35	60	40	2.65	2.70	VI

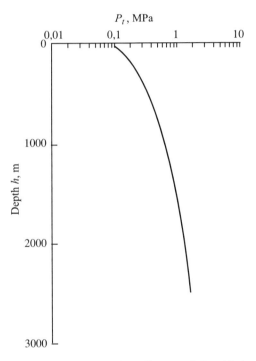

Fig. 10.4. A curve showing the change of tensile strength P_t with depth of occurrence h for clay seals of classes I–II calculated using the «Seal Test» program.

For clay seals of classes III–IV:

$$P_t = 0.0915 + 5.89334 \times 10^{-4} \times h + 6.01741 \times 10^{-8} \times h^2; \quad R = 0.99$$

$$(10.5)$$

For clay seals of classes V–VII:

$$P_t = -0.17364 + 0.00103 \times h - 3.026 \times 10^{-7} \times h^2; \quad R = 0.98 \quad (10.6)$$

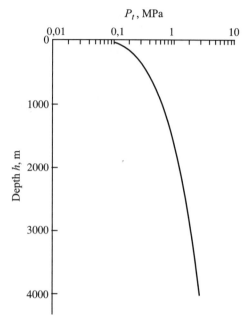

Fig. 10.5. A curve showing the change of tensile strength P_t with depth of occurrence h for clay seals of classes III–IV calculated using the «Seal Test» program.

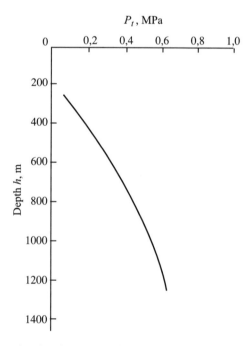

Fig. 10.6. A curve showing the change of tensile strength P_t with depth of occurrence h for clay seals of classes V–VII calculated using the «Seal Test» program.

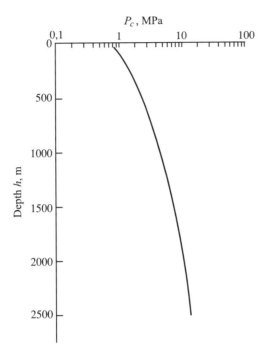

Fig. 10.7. A curve showing the change of compression strength P_c with depth of occurrence h for clay seals of classes I–II calculated using the «Seal Test» program.

Further, based on the data on P_t and using the «Seal Test» program, the compressive strength was also calculated for clay seals of different classes. The results of such a calculation of P_c change with depth are given in Figures 10.7–10.9. They show that the $P_c = f(h)$ dependence for all classes of clay seals, as the $P_t = f(h)$ dependence, follows the quadric polynomial equation:

1. For clay seals of classes I–II:

$$P_c = 0.69725 + 0.00282 \times h + 1.09787 \times 10^{-6} \times h^2; \quad R + 0.98 \quad (10.7)$$

2. For clay seals of classes III–IV:

$$P_c = 0.41061 + 0.005 \times h + 9.00068 \times 10^{-7} \times h^2; \quad R = 0.99 \quad (10.8)$$

3. For clay seals of classes V–VII:

$$P_c = -0.12727 + 0.00386 \times h + 1.17542 \times 10^{-6} \times h^2; \quad R = 0.99$$
$$(10.9)$$

Figures 10.10–10.12 show the results of comparison of the experimental curves $P_c = f(h)$ with the theoretical curves calculated using the «Seal Test» program for clay seals of classes I–II, III–IV, and V–VII, respectively.

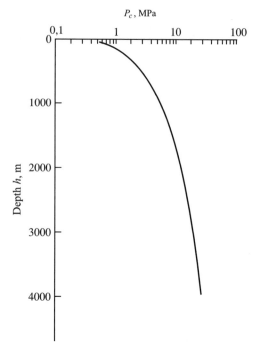

Fig. 10.8. A curve showing the change of compression strength P_c with depth of occurrence h for clay seals of classes III–IV calculated using the «Seal Test» program.

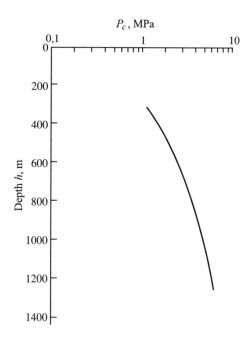

Fig. 10.9. A curve showing the change of compression strength P_c with depth of occurrence h for clay seals of classes V–VII calculated using the «Seal Test» program.

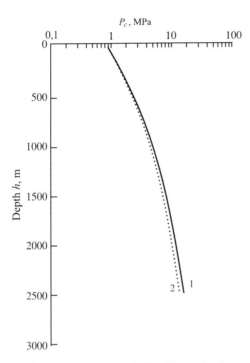

Fig. 10.10. Change of compression strength P_c with depth of occurrence h for clay seals of classes I–II: 1 – dependence calculated using the «Seal Test» program; 2 – experimental dependence.

An analysis of the obtained results shows that in the depth intervals, where clay seals have high screening properties, a good agreement is observed between the theoretical and the experimental data. It should be noted, that the experimental curves pass a little lower than the theoretical ones.

Such a trend of the experimental dependence $P_c = f(h)$ may be explained by some decompaction of clay rock samples during the recovery of core from great depths to the surface. As a result, their strength proves to be less than the theoretically calculated values.

The best agreement between the theoretical and the experimental data is registered for the clay seals of classes I–II and V–VII within the entire range of considered depths (from ~120 to 2500 m and from 450 to 1200 m, respectively) (Figs. 10.10 and 10.12). It should be noted, that clay seals of classes I–II and V–VII have high screening properties at these depths.

The greatest differences between the theoretical and the experimental data are observed in clay seals of classes III–IV at depths exceeding 2500 m (Fig. 10.11, a dotted portion of curve 1), when they begin to lose their screening properties. As already noted earlier, this fact is explained by the development of fracturing processes at great depths and by the rise of abundant microfractures in the samples. As the theoretical calculation does not take into account the reduction of strength due to the rise of microfractures in rocks, the experimental values of P_c prove to be lower than the theoretically expected values.

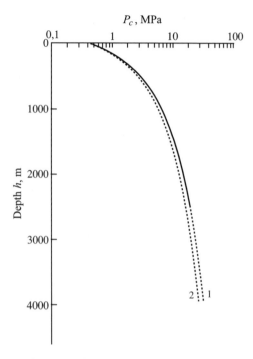

Fig. 10.11. Change of compression strength P_c with depth of occurrence h for clay seals of classes III–IV: 1 – dependence calculated using the «Seal Test» program; 2 – experimental dependence.

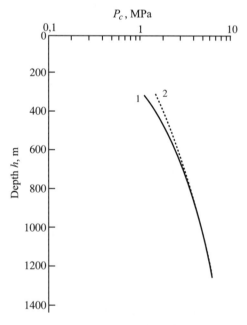

Fig. 10.12. Change of compression strength P_c with depth of occurrence h for clay seals of classes V–VII: 1 – dependence calculated using the «Seal Test» program; 2 – experimental.

In conclusion, it should be noted, that for certain depth intervals, where clay seals have high screening properties, the data on the mechanical strength of clay seals obtained using the «Seal Test» program give a good agreement with the experimental results. Thus, for clay seals of classes I–II, such an approximate calculation of P_c proves to be correct at depths ranging from ~500 to 6000 m, for clay seals of classes III–IV – from ~1000 to 2500 m, and for clay seals of classes V–VII – from ~1200 to 1800 m.

References

1. Vassoevich, N.B.: The Role of Clays in the Origin of Oil. *Sovetskaya Geologiya* No. 3 (1975), pp. 15–29 (in Russian).
2. Pettijohn, F.J.: *Sedimentary Rocks (3rd ed.)*. Harper & Row Publishers, New York, Evanston, San Francisco and London, 1975.
3. Sergeev, E.M., Golodkovskaya, G.A., Ziangirov, R.S., Osipov, V.I., Trofimov, V.T.: *Soil Science (4th ed.)*. Izd. Mosk. Univ., Moscow, 1983 (in Russian).
4. Deryagin, B.V.: The Theory of Slow Coagulation and Stability of Weakly Charged Lyophobic Sols and Emulsions. *Kolloidnyy Zhurnal* 7, No. 3 (1941), pp. 285–287 (in Russian).
5. Deryagin, B.V., Landau, L.D.: The Theory of Stability of Strongly Charged Particles in Electrolyte Solutions. *ZhEhTF* 15. No. 11 (1945), pp. 663–682 (in Russian).
6. Deryagin, B.V.: *The Theory of Stability of Colloids and Thin Films*. Nauka, Moscow, 1986 (in Russian).
7. Osipov, V.I., Sergeev, E.M.: Crystallochemistry of Clay Minerals and Their Properties. In: E.M. Sergeev (ed.): *Engineering-Geological Properties of Clay Rocks and the Processes in Them*. No. 1. Izd. Mosk. Univ., Moscow, 1972, pp. 13–26 (in Russian).
8. Zlochevskaya, R.I.: *Bound Water in Clay Soils*. Izd. Mosk. Univ., Moscow, 1969 (in Russian).
9. Kul'chitskiy, L.I.: *The Role of Water in the Formation of Clay Rock Properties*. Nedra, Moscow, 1975 (in Russian).
10. Shchukin, E.D. (ed.): *Surface Films of Water in Fine Grained Structures*. Izd. Mosk. Univ., Moscow, 1988 (in Russian).
11. Green-Kelly, D., Deryagin, B.V.: Double Beam Refraction of thin Liquid Layers. *Doklady AN SSSR* 153, No. 3 (1963), pp. 638–641 (in Russian).
12. Deryagin, B.V., Churaev, N.V.: *The Wetting Films*. Nauka, Moscow, 1984 (in Russian).
13. Kiselev, V.F., Kvlividze, V.I., Zlochevskaya, R.I. (eds.): *Bound Water in Fine Grained Systems*. No. 5. Izd. Mosk. Univ., Moscow, 1980 (in Russian).
14. Bondarenko, N.F.: *The Physics of Underground Water Movement*. Gidrometizdat, Leningrad. 1973 (in Russian).
15. Zavaritskiy, A.G.: *Igneous Rocks*. Izd. AN SSSR, Moscow, 1955 (in Russian).
16. Frolov, V.T.: *Lithology*. Book 1, Izd. Mosk. Univ., Moscow, 1922 (in Russian).
17. Rebinder, P.A.: Physicochemical Mechanics of Dispersed Structures. In: P.A. Rebinder (ed.): *Physicochemical Mechanics of Dispersed Structures*. Nauka, Moscow, 1966, pp. 3–16 (in Russian).

18. Osipov, V.I.: The Concept of «Soil Structure» in Engineering Geology. *Inzhenernaya Geologiya* No. 3 (1985), pp. 4–18 (in Russian).

19. Rekshinskaya, L.G.: *Atlas of Electron Microphotographs of Clay Minerals and Their Natural Associations in Sedimentary Rocks*. Nedra, Moscow, 1966 (in Russian).

20. Beutelspacher, H., van der Marel, H.W.: *Atlas of Electron Microscopy of Clay Minerals and Their Admixtures*. Elsevier Publishing Company, Amsterdam, London, New York, 1968.

21. Grabowska-Olszewska, B., Osipov, V., Sokolov, V.: *Atlas of the Microstructure of Clay Soils*. Warszawa, Panstwowe Widawnictwo Naukowe, 1984.

22. Sergeev, E.M. (ed.): *Methodical Handbook on the Engineering-Geological Study of Rocks*. 2, Nedra, Moscow, 1984 (in Russian).

23. Sokolov, V.N.: Engineering-geological classification of clay microstructures. D.G. Price (ed.): *Proc. Sixth Int. Congress International Association of Engineering Geology*. A.A. Balkema, Rotterdam-Brookfield, 1990, pp. 753–760.

24. Verwey, E., Overbeek, T.: *Theory of the Stability of Lyophobic Colloids*. Elsevier Publ., New York, Amsterdam, 1948.

25. Deryagin, B.V., Obukhov, E.V.: Anomalous Properties of Thin Layers of Liquids. III. Ultramicrometric Investigations of Lyospheres (Solvate Envelopes) and an «Elementary» Act of Swelling. *Kolloidnyy Zhurnal* 1, No. 5 (1935), pp. 385–398 (in Russian).

26. Deryagin, B.V., Churaev, N.V.: The Isotherm of the Wedging Pressure of Water Films on the Quartz Surface. *Doklady AN SSSR* 207, No. 3 (1972), pp. 572–575 (in Russian).

27. Deryagin, B.V., Zorin Z.M.: Investigation of the Surfaces of Vapor Condensation and Adsorption in the Vicinity of Saturation Using the Optical Micropolarization Method. *Zhurnal Fiz. Khimii* 29, No. 10 (1955), pp. 1755–1770 (in Russian).

28. Israelachvili, J.N., Adams, G.E.: Measurement of Forces between Two Mica Surfaces in Aqueous Solution. *Journ. Chem. Soc. Faraday Trans.* P.1, 74, No. 4 (1978), pp. 975–1001.

29. Pashley, R.M.: DLVO and Hydration Forces between Mica Surfaces in Li^+, Na^+, K^+, and Cs^+ Electrolyte Solutions. A Correction of Double-Layer and Hydration Forces with Surface Cation Exchange Properties. *J. Colloid and Interface Sci.* 83, No. 2 (1981), pp. 531–546.

30. Israelachvili, J.N., Pashley, R.M.: Molecular Layering of Water at Surface and Origin of Repulsive Hydration Forces. *Nature* 306, No. 5940 (1983), pp. 249–250.

31. Marčelja, S., Radič, N.: Repulsion of Interfaces due to Boundary Water. *Chem. Phys. Lett.* 42, No. 1 (1976), pp. 129–134.

32. Shchukin, E.D., Pertsov, A.V., Amelina, E.A.: *Colloid Chemistry*. Izd. Mosk. Univ., Moscow, 1982 (in Russian).

33. London, F.: The General Theory of Molecular Forces. *Trans. Faraday Soc.* 33, No. 8 (1937), pp. 8–26.

34. Hamaker, H.C.: The London–van der Waals Attraction between Spherical Particles. *Physica* 4, No. 10 (1937), p. 1058.

35. Lifshits, E.M.: The Theory of Molecular Attraction Forces between Condensed Bodies. *Doklady AN SSSR* 97, No. 4 (1954), pp. 643–647 (in Russian).

36. Lifshits, E.M.: The Theory of Molecular Attraction Forces between Solids. *Zhurnal Ehksper. i Teor. Fiziki* 29, No. 1 (1955), pp. 94–110 (in Russian).

37. Kruyt, H.R.: *The Science of Colloids* 1. Izd. Inostr. Lit., Moscow, 1955 (in Russian).

38. Yaminskiy, V.V., Pchelin, V.A., Amelina, E.A., Shchukin, E.D.: *Coagulation Contacts in Dispersed Systems*. Khimiya, Moscow, 1982 (in Russian).

39. Efremov, I.F.: *Periodical Colloidal Structures*. Khimiya, Leningrad, 1971 (in Russian).

40. Sonntag, G., Strenge, K.: *Coagulation and Stability of Dispersed Systems*. Khimiya, Leningrad, 1973 (in Russian).

41. Deryagin, B.V., Churaev, N.V., Muller, V.M.: *Surface Forces*. Nauka, Moscow, 1985 (in Russian).

42. Shchukin, E.D.: On Some Problems in the Physicochemical Theory of the Strength of Finely Dispersed Porous Bodies – Catalysts and Sorbing Agents. *Kinetika i Kataliz* 6, No. 4 (1965), pp. 641–650 (in Russian).

43. Shchukin, E.D., Amelina, E.A., Yusupov, R.K., Rebinder, P.A.: Experimental Investigation of Cohesion Forces in Individual Microscopic Contacts between Crystals in Squeezing-Out and Caking. *Kolloidnyy Zhurnal* 31, No. 6 (1969), pp. 913–918 (in Russian).

44. Fuks, G.I.: On the Forces of Contact Interactions of Solid Particles in a Liquid Medium. In: P.A. Rebinder, G.I. Fuks (eds.): *The Progress in Colloid Chemistry*. Nauka, Moscow, 1973. pp. 117–129 (in Russian).

45. Polak, A.F.: On the Mechanics of Structure Formation in the Hardening of Monomineral Astringents. *Kollodnyy Zhurnal* 24, No. 2 (1962), pp. 206–214 (in Russian).

46. Polak, A.F., Fazullin, I.Sh.: Problems in the Physicochemical Mechanics of Clay Soils. *Kolloidnyy Zhurnal* 33, No. 2 (1971). pp. 258–263 (in Russian).

47. Yakhnin, E.D., Taubman, A.B.: On the Problem of Structure Formation in Dispersed Systems. *Doklady AN SSSR* 155, No. 1 (1964), pp. 179–184 (in Russian).

48. Gor'kova, I.M.: Clay Rocks and Their Strength in the Light of Modern Concepts of Colloid Chemistry. *Trudy Laborat. Gidrogeol. Problem* 15 (1957), pp. 26–52 (in Russian).

49. Korobanova, I.G.: *Formation of Engineering–Geological Properties of a Terrigenous Sequence (with the Mesozoic of the Russian Platform as an Example)*. Nauka, Moscow, 1970 (in Russian).

50. Osipov, V.I.: Engineering-Geological Classification of Clay Rocks. *Vestnik Mosk. Univ., Ser. Geol.*, No. 4 (1976), pp. 17–30 (in Russian).

51. Osipov, V.I.: *The Nature of the Strength and Deformation Properties of Clay Rocks*. Izd. Mosk. Univ., Moscow, 1976 (in Russian).

52. Osipov, V.I., Sokolov, V.N.: The Role of Ionic–Electrostatic Forces in the Formation of Structural Bonds in Clays. *Vestnik Mosk. Univ., Ser. Geol.* No. 1 (1974), pp. 16–32 (in Russian).

53. Osipov, V.I., Sokolov, V.N.: Structural Bonds and the Processes of Structure Formation in Soils. In: E.M. Sergeev (ed.): *Theoretical Principles of Engineering Geology. The Physicochemical Principles.* Nedra, Moscow, 1985, pp. 104–105 (in Russian).

54. Sokolov, V.N.: *Investigation of Structural Bond Formation in Clays during Their Dehydration.* Candidate Sci. Thesis (Author's abstract). Moscow State Univ., Moscow, 1973 (in Russian).

55. Sokolov, V.N.: The Influence of Moisture on the Strength of Structural bonds in Clay Particles. *Vestnik Mosk. Univ., Ser. Geol.* No. 6 (1973), pp. 100–104 (in Russian).

56. Kul'chitskiy, L.I., Us'yarov, O.G.: *Physicochemical Principles of the Formation of Clay Rock Properties.* Nedra, Moscow, 1981 (in Russian).

57. Osipov, V.I.: Structure Bonds and the Properties of Clays. *Bull. of the IAEG* No. 12 (1975), pp. 13–20.

58. Deryagin, B.V., Krotova, N.A., Smilga, V.P.: *Adhesion of Solids.* Nauka, Moscow, 1975 (in Russian).

59. Schofield, R.K., Samson, H.R.: Flocculation of Kaolinite due to the Attraction of Oppositely Charged Crystal Faces. *Disc. Faraday Soc.* No. 18 (1954), pp. 135–145.

60. Tolstoy, N.A., Spartakov, A.A., Trusov, A.A.: Strong Electric Dipole Moment of Colloidal Particles. In: B.V. Deryagin (ed.): *Investigations in the Field of Surface Forces.* Nauka, Moscow, 1967, pp. 56–78 (in Russian).

61. Osipov, Yu.B.: *Investigation of Clay Suspensions, Pastes, and Sediments in the Magnetic Field.* Izd. Mosk. Univ., Moscow, 1968 (in Russian).

62. Deryagin, B.V., Abrikosova, I.I.: Direct Measurement of Molecular Attraction as a Function of Distance between Surfaces. *ZhEhTF* 21, No. 8 (1951), pp. 1755–1770 (in Russian).

63. Rabinovich, Ya.I.: Direct Measurements of the Wedging Pressure in Electrolyte Solutions as a Function of Distance between Crossed Threads. *Kolloidnyy Zhurnal* 39, No. 6 (1977), pp. 1094–1100 (in Russian).

64. Babak, V.G., Sokolov, V.N., Sveshnikova, E.V., Osipov, V.I.: Direct Measurement of Cohesion Forces between Mica Particles in Aqueous Solutions of Polyvinyl Alcohol. *Inzhenernaya Geologiya* No. 4 (1984), pp. 57–63 (in Russian).

65. Push, R.: Structural Variations in Boulder Clay. In: L. Barden, R. Push (eds.): *Proc. Int. Symp. Soil Structure. Gothenburg.* Swedish Geotechnical Society, Stockholm, 1973, pp. 113–121.

66. Smart, P.: Contribution to Stress-Strain Behavior of Soil. R.H.G. Parry (ed.): *Proc. Roscoe Memorial Symp. Cambridge Univ.*, G.T. Foulis Co. Ltd, Honley on Thames, 1971. pp. 253–255.

67. Osipov, V.I., Sokolov, V.N.: Relation between the Fabric of Clay Soils and Their Origin and Degree of Compaction. *Bull. of the IAEG* No. 18 (1978), pp. 73–81.

68. Osipov, V.I., Sokolov, V.N.: Structure Formation in Clay Sediments. *Bull. of the IAEG* No. 18 (1978) pp. 83–90.

69. Osipov, V.I., Sokolov, V.N., Rumyantseva, N.A.: *Microstructure of Clay Rocks.* Nedra, Moscow, 1989 (in Russian).

70. O'Brien, N.R.: Fabric of Kaolinite and Illite Floccules. *J. Clays and Clay Minerals* 19 (1971), pp. 353–359.

71. Push, R.: Clay Microstructure – a Study of Microstructure of Soft Clays with Special Reference to their Physical Properties. *St. Inst. Byggnadsforsk. Docum. D8.* Svensk Byggtjänst, Stockholm, 1970, pp. 93–101.

72. Lombard, A.: *Géologie Sédimentaire*. Masson, Paris, 1956.

73. Millot, G.: *Géologie des Argiles (Alternations, Sedimentologie, Géochimie)*. Masson et Cie, editeurs 120, boulevard Saint-Germain, Paris-VIe, 1964.

74. Fersman, A.E.: *Geochemistry of Russia*. ONTI, Leningrad, 1922 (in Russian).

75. Kossovskaya, A.G., Logvinenko, N.V., Shutov, V.D.: On the Stages in the Formation and Alteration of Terrigenous Rocks. *Doklady AN SSSR* 116, No. 2 (1957), pp. 293–296 (in Russian).

76. Vassoevich, N.B.: Once More on the Terminology Used to Designate the Stages and Substages of Lithogenesis. *Trudy VNIGRI* 190 (1962), pp. 242–243 (in Russian).

77. Strakhov, N.M.: *The Principles of the Lithogenesis Theory.* Vol. 1. Izd. AN SSSR, Moscow, 1960 (in Russian).

78. Strakhov, N.M. *The Principles of the Lithogenesis Theory.* Vols. 2, 3. Izd. AN SSSR, Moscow, 1962 (in Russian).

79. Logvinenko, N.V., Orlova, L.V.: *Formation and Alteration of Sedimentary Rocks on the Continent and in the Ocean.* Nedra, Leningrad, 1987 (in Russian).

80. Vassoevich, N.B.: Stages of Lithogenesis. In: N.B. Vassoevich (ed.): *Handbook on Lithology.* Nedra, Moscow, 1983. pp. 85–96 (in Russian).

81. Logvinenko, N.V.: *Postdiagenetic Alterations of Sedimentary Rocks.* Nedra, Leningrad, 1968 (in Russian).

82. Shutov, V.D.: *Mineral Paragens of Graywacke Complexes.* Nauka, Moscow. 1975 (in Russian).

83. Fife, W.S., Price, N.J., Tompson, A.B.: *Fluids in the Earth's Crust.* Elsevier Scientific Publishing Company, Amsterdam, Oxford, New York, 1978.

84. Pustovalov, L.V.: Secondary Alterations of Sedimentary Rocks and Their Geological Implications. In: L.V. Pustovalov (ed.): *On the Secondary Alterations of Sedimentary Rocks.* Izd. AN SSSR, Moscow, 1956, pp. 3–52 (in Russian).

85. Kopeliovich, A.V.: *Epigenesis of Ancient Sequences in the South-West of the Russian Platform.* Nauka, Moscow, 1965 (in Russian).

86. Rukhin, L.B.: *The Principles of Lithology.* Nedra, Leningrad, 1969 (in Russian).

87. Strakhov, N.M.: On the Theoretical Lithology and its Problems. *Izvestiya AN SSSR, Ser. Geol.* No. 11 (1957), pp. 15–31 (in Russian).

88. Strakhov, N.M., Logvinenko, N.V.: On the Stages of Sedimentary Rock Formation and Their Designation. *Doklady AN SSSR* 125, No. 2 (1959), pp. 389–392 (in Russian).

89. Müller, G.: Diagenesis and Catagenesis of Clay Sediments. In: G. Larsen, G.V. Chilingar (eds.): *Diagenesis in Sediments.* Elsevier Publishing Company, Amsterdam, London, New York, 1967 pp. 122–164.

90. Meade, R.H.: Factors Influencing the Early Stages of the Compaction of Clays and Sands. Review. *Journ. Sedimentary Petrol.* 36, No. 4 (1966), pp. 1085–1101.

91. Sergeev, E.M., Grabowska-Olszewska, B., Osipov, V.I., Sokolov, V.N.: Types of Clay Rock Microstructures. *Inzhenernaya Geologiya* No. 2 (1979), pp. 48–58 (in Russian).

92. Sergeev, E.M. (ed.): *Theoretical Principles of Engineering Geology. The Geological Principles.* Nedra, Moscow, 1985 (in Russian).

93. Sergeev, E.M. (ed.): *Theoretical Principles of Engineering Geology. The Physicochemical Principles.* Nedra, Moscow, 1985 (in Russian).

94. Sergeev, E.M.: On the Problem of Silty Soil Compaction by Heavy Loading. *Vestnik Mosk. Univ.* No. 1 (1946), pp. 91–93 (in Russian).

95. Lomtadze, V.D.: On the Formation of Clay Rock Properties. *Zapiski Leningr. Gorn. Inst.* 32, No. 2 (1956), pp. 41–87 (in Russian).

96. Mukhin, Yu.V.: *The Processes of Clay Sediment Compaction.* Nedra, Moscow, 1963 (in Russian).

97. Blokh, A.M.: *The Structure of Water in Geological Processes.* Nedra, Moscow, 1969 (in Russian).

98. Tarasevich, Yu.I., Ovcharenko, F.D.: *Adsorption on Clay Minerals.* Naukova Dumka, Kiev, 1973 (in Russian).

99. Weaver, C.E.: Possible Uses of Clay Minerals in the Search for Oil. *Bull. AAPG* 44, No. 9 (1960), pp. 1505–1519.

100. Burst, J.F.: Postdiagenetic Clay Mineral Environment Relationships in the Gulf Coast Eocene. *J. Clays and Clay Minerals* 6 (1959), pp. 327–341.

101. Power, M.C.: Fluid-Release Mechanism in Compacting Marine Mudrocks and Their Importance in Oil Exploration. *Bull. AAPG* 51, No. 7 (1967), pp. 1240–1254.

102. Kossovskaya, A.G.: Mineralogy of the Mesozoic Terrigenous Complex of the Vilyuy Depression and Western Verkhoyansk Region. *Trudy GIN AN SSSR* No. 63. Izd. AN SSSR, Moscow (1962), pp. 24–36 (in Russian).

103. Zkhus, I.D., Bakhtin, V.V.: *Lithogenetic Transformations of Clays in the Zone of Anomalously High Formation Pressures.* Nauka, Moscow, 1979 (in Russian).

104. Frank-Kamenetskiy, V.A. (ed.): *Radiography of Main Types of Rock Building Minerals.* Nedra, Leningrad, 1983 (in Russian).

105. Chikhradze, G.A.: Lithology of the Lower and Middle Jurassic Sediments on the Southern Flank of the Great Caucasus. *Trudy GIN AN GSSR* No. 62. Metsniereba, Tbilisi (1979), pp. 177–189 (in Russian).

106. Pashaly, N.V., Kheirov, M.B.: Clay Minerals from the Productive and Red-Bed Formations of the Shallow-Water Zone in the Southern Caspian Sea. *Litologiya i Poleznye Iskopaemye* No. 5 (1975), pp. 19–29 (in Russian).

107. Prozorovich, G.Eh.: Seals for Oil and Gas Pools. In: P.K. Kulikov (ed.): *Trudy Zap.-Sib. NIGRI* No. 49, Nedra, Moscow, 1972 (in Russian).

108. Vassoevich, N.B.: The Theory of Sedimentary-Migrational Origin of Oil (Historical Review and Present State). *Izvestiya AN SSSR, Ser. Geol.* No. 11 (1967), pp. 135–156 (in Russian).

109. Kosygin, Yu.A., Varnavskiy, V.G. (eds.): *Lithological-Petrographical Criteria for Oil and Gas Occurrence*. Nauka, Moscow, 1990 (in Russian).

110. Landau, L.D.: *The Theory of Elasticity*. Nauka, Moscow, 1965 (in Russian).

111. Yusupov, R.K.: *Physicochemical Investigations of the Formation Conditions and Strength of Microscopic Phase Contacts between Individual Solid Particles*. Candidate Sci. Thesis (Author's abstract). Moscow State Univ., Moscow, 1973 (in Russian).

112. Akhmatov, A.S.: *Molecular Physics of Boundary Friction*. Fizmatgiz, Moscow, 1963 (in Russian).

113. Geguzin, Ya.E.: *Physics of Caking*. Nauka, Moscow, 1967 (in Russian).

114. Rebinder, P.A., Segalova, E.E.: New Problems in the Colloid Chemistry of Mineral Astringents. *Priroda* No. 12 (1952), pp. 306–308 (in Russian).

115. Rebinder, P.A.: Physicochemical Concepts about the Mechanism of the Mineral Astringent Substance Setting and Hardening Processes. In: P.P. Budnikov, Y.M. Butt, S.M. Royak, M.O. Yushkevich (eds.): *Proc. Conf. on the Chemistry of Cements*. Promstroyizdat, M. 1956 (in Russian).

116. Mikhaylov, N.V.: *The Main Principles of New Concrete and Ferro-Concrete Technology*. Stroyizdat, Moscow, 1961 (in Russian).

117. Polak, A.F., Babkov, V.V., Andreeva, E.P.: *Solidification of Mineral Astringent Substances*. Bashkirskoye Knizhnoye Izd., Ufa, 1990 (in Russian).

118. Lyubimova, T.Y.: Peculiarities of the Crystallization Hardening of Mineral Astringents in the Zone of Contact with Different Solids (Fillers). In: P.A. Rebinder (ed.): *Physicochemical Mechanics of Dispersed Structures*. Nauka, Moscow, 1966, pp. 268–280 (in Russian).

119. Ur'ev, N.B., Mikhaylov, N.V.: Peculiarities of the Structure Formation Processes in thin Interlayers of the Cement-Water Suspensions (Colloidal-Cement Glue). In: P.A. Rebinder (ed.): *Physicochemical Mechanics of Dispersed Structures*. Nauka, Moscow, 1966, pp. 290–296 (in Russian).

120. Rebinder, P.A., 1968. Problems in the Formation of Dispersed Systems and Structures in these Systems: the Physicochemical Mechanics of Dispersed Structures and Solids. In: P.A. Rebinder (ed.): *Modern Problems in Physical Chemistry*. Vol. 3. Izd. Mosk. Univ., Moscow, pp. 334–414 (in Russian).

121. Gor'kova, I.M.: *Structural and Deformational Features of Sedimentary Rocks with Different Degrees of Compaction and Lithification*. Nauka, Moscow, 1965 (in Russian).

122. Bocharova, I.S.: On the Problem of the Change in the Composition and Structure-Forming Properties of Organic Matter in Clay Rocks in the Process of Lithogenesis. In: I.V. Popov (ed.): *Formation of the Engineering–Geological Properties of Clay Rocks in the Process of Lithogenesis*. Izd. AN SSSR, Moscow, 1963 (in Russian).

123. Korobanova, I.G.: *Regularities in the Formation of Terrigenous Sediment Properties*. Nauka, Moscow, 1983 (in Russian).

124. Larsen, G., Chilingar, G.V.. (eds.): *Diagenesis in Sediments*. Elsevier Publishing Company, Amsterdam, London, New York, 1967.

125. Aliev, F.S.: On the Problem of Diagenesis of Clay Rocks from the Baku Archipelago. *Doklady AN SSSR* 127, No. 6 (1959), pp. 1263–1264 (in Russian).

126. Polyakov, A.S.: On the Post-Sedimentary Transformation of Late Cenozoic Clay Sediments in the Present Interior Seas. In: P.P. Timofeev (ed.): *Stratigraphy and Lithology of the Mesozoic-Cenozoic Sedimentary Cover of the World Ocean. Pap. Abstr.* Vol. 2. GIN, Moscow, 1984. pp. 77–79 (in Russian).

127. Ziangirov, R.S.: *The Nature of Clay Soil Compressibility.* Dr. Sc. (Geol.-Min.) Thesis (Author's abstract). Moscow State Univ., Moscow, 1974 (in Russian).

128. Emery, K.O., Rittenberg, S.C.: Early Diagenesis of California Basin Sediments in Relation to Origin of Oil. *Bull. AAPG* 36, No. 5 (1952), pp. 735–806.

129. Chilingar, G.V., Knight, L.: Relationship between Pressure and Moisture Content of Kaolinite, Illite and Montmorillonite Clay. *Bull. AAPG* 44, No. 1 (1960), pp. 101–106.

130. Weller, J.: Compaction of Sediments. In: N.B. Vassoevich (ed.): *Problems of Petroleum Geology in the Interpretation of Foreign Scientists.* Gostopizdat, Leningrad, 1961, pp. 84–137 (in Russian).

131. Meade, R.H.: *Removal of Water and Rearrangement of Particles during the Compaction of Clayey Sediments. Review.* U.S. Geological Survey Professional Papers 497-B, U.S. Government Printing Office, 1964.

132. Vassoevich, N.B.: An essay of constructing a gravity compaction curve for clay sediments. *Novosti Neftyanoy i Gazovoy Tekhniki. Ser. Geol.* 4 (1960), pp. 11–15 (in Russian).

133. Prozorovich, Eh.A.: Compaction of Clay Rocks of the Maykopian Formation with the Growth of Their Occurrence Depth in Some Regions of the Caucasus. In: N.I. Shapirovskiy (ed.): *Scientific-Technical Information of the Azerbaijanian Scientific-Research Inst. for Petrol. Prod.* No. 3, AzINTI, Baku, 1961, pp. 57–61 (in Russian).

134. Yapaskurt, O.V.: *Stage Analysis of Lithogenesis. A Textbook.* Izd. Mosk. Univ., Moscow, 1995 (in Russian).

135. Korobanova, I.G., Kovaleva, A.P., Kopylov, A.K.: Composition and Physical-Mechanical Properties of Clay Sediments and Their Alterations in the Process of Lithogenesis. In: I.V. Popov (ed.): *Engineering-Geological Properties of Rocks and the Problems in Lithogenesis.* Nauka, Moscow, 1965, pp. 12–33 (in Russian).

136. Khanin, A.A., Abdurakhimov, K.A., Lazareva, V.N.: Properties of Clay Seals from the Mesozoic-Cenozoic Oil- and Gas-Bearing Sedimentary Complex of Some Areas in the Pre-Caucasus Region. *Geologiya Nefti i Gaza* No. 10 (1969), pp. 47–50 (in Russian).

137. Lazareva, V.M.: On the Problem of Studying the Influence of Rock Composition on the Change of Their Physical Properties. In: V.A. Frank-Kamenetskiy (ed.): *Materials of the XI All-Union Conference on the Study and Use of Clays and Clay Minerals.* Izd. IGEM, Moscow, 1976, pp. 171–173 (in Russian).

138. Lazareva, V.M.: Peculiarities of Clay Seals of the Fergana Depression and Estimation of Their Isolating Properties. In: V.A. Frank-Kamenetskiy (ed.): *Materials of the XI All-Union Conference on the Study and Use of Clays and Clay Minerals*. Izd. IGEM, Moscow, 1976, pp. 173–175 (in Russian).

139. Martirosova, O.A., Bakhtin, V.V.: Regularities in the Compaction of Maykopian Clays of Azerbaijan with Depth and the Influence of Over-pressures on the Compaction Process. In: V.A. Frank-Kamenetskiy (ed.): *Materials of the XI All-Union Conference on the Study and Use of Clays and Clay Minerals*. Izd. IGEM, Moscow, 1976, pp. 178–179 (in Russian).

140. Khanin, A.A.: *Reservoir Rocks for Oil and Gas*. Nedra, Moscow, 1969 (in Russian).

141. Vasil'ev, Yu. (ed.): *Aralsor Ultradeep Well (Geological Results)*. Trudy MINKh i GP, No. 10, Nedra, Moscow, 1972 (in Russian).

142. Marmorshtein, L.M.: *The Reservoir and Screening Properties of Sedimentary Rocks under Various Thermodynamic Conditions*. Nedra, Leningrad, 1975 (in Russian).

143. Strakhov, N.M.: On the Understanding of Diagenesis. In: N.M. Strakhov (ed.): *Problems in the Mineralogy of Sedimentary Deposits*. Izd. L'vov. Univ., L'vov, pp. 3–6, 1956 (in Russian).

144. Engel'gardt, V.: *Pore Space of Sedimentary Rocks*. Nedra, Moscow, 1964 (in Russian).

145. Polyakov, A.S., Osipov, V.I., Kotlov, V.F., Kuprin, P.N.: Change in the Microstructure and Physical-Mechanical Properties of Clay Sediments in the Baku Archipelago with Passage from the Diagenesis to the Catagenesis stage. *Inzhenernaya Geologiya* No. 5 (1979), pp. 29–40 (in Russian).

146. Chillingarian, G.V., Rieke, H.H.: *Compaction of Argillaceous Sediments*. Elsevier Scientific Publishing Company, Amsterdam, 1973.

147. Hager, J., Handin, J.: Experimental Deformation of Sedimentary Rocks under Confining Pressure Tests at Room Temperature on Dry Samples. *Bull. AAPG* 41, No. 1 (1957), pp. 41–53.

148. Vassoevich, N.B.: Oil Formation in Terrigenous Sediments. *Trudy. VNIGRI* 128 (1958), pp. 9–220 (in Russian).

149. Goryunov, Yu.V, Pertsev, N.V., Summ, B.D.: *Rebinder's Effect*. Nauka, Moscow, 1966 (in Russian).

150. Gol'dberg, V.M., Skvortsov, N.P.: *Permeability and Filtration in Clays*. Nedra, Moscow, 1986 (in Russian).

151. Purcell, W.R.: Capillary Pressures – Their Measurement Using Mercury and the Calculation of Permeability Therefrom. *Trans. AIME* 186, TP 2663 (1949), pp. 39–46.

152. Burdine, N.T., Gournary, L.S., Richerts, P.P.: Pore Size Distribution of Petroleum Reservoir Rocks. *J. Petrol. Technol* 2, No. 7 (1950), pp. 195–204.

153. Briling, I.A.: 1988. Water Movement in Soils and Clay Rocks. In: E.D. Shchukin (ed.): *Surface Films of Water in Dispersion Structures*. Izd. Mosk. Univ., Moscow, pp. 74–90 (in Russian).

154. Berezkina, G.M., Briling, I.A., Koryakina, N.S., Krasnushkin, A.V.: Experimental Investigations on the Movement of Moisture and Salts in Clay Rocks. In: Yu.O. Zeegofer (ed.): *Interaction of Surface and Subsurface Drainage*. Izd. Mosk. Univ., Moscow, No. 4, 1976, pp. 196–270 (in Russian).

155. Babinets, A.E., Emel'anov, V.A., Mitropol'skiy, A.Y.: *Physicochemical Properties of Bottom Sediments of the Black Sea*. Naukova Dumka, Kiev. 1981 (in Russian).

156. Bakhtin, V.V., Fomin, A.A., Kuz'menkova, G.E.: Lithophysical and Deformation Properties of Visean Seals in the Dniepr-Donets Trough under Subsurface Conditions. In: M.P. Volarovich (ed.): *Physical Properties of Oil Reservoirs under High Pressures and Temperatures*. Nauka, Moscow, 1979, pp. 77–91 (in Russian).

157. Kuz'menkova, G.E., Fomin, A.A.: Isolating Property, Deformational and Strength Characteristics of Clay Rocks. In: M.P. Volarovich (ed.): *Physical Properties of Reservoir Rocks for Oil at High Pressures and Temperatures*. Nauka, Moscow, 1979, pp. 5–19 (in Russian).

158. Gor'kova, I.M.: *Theoretical Principles of Sedimentary Rock Estimation with Engineering-Geological Purposes*. Nauka, Moscow, 1966 (in Russian).

159. Rabaev, G.S.: *Uniaxial Compressibility of Dispersed Soils and the Factors that Determine It*. Candidate Sci. Thesis (Author's abstract). Moscow State Univ., Moscow, 1969 (in Russian).

160. Volarovich, M.P.: Investigation of Physical-Mechanical Properties of Rocks under High Pressures and Temperatures. In: N.I. Khitarov (ed.): *Experimental Investigations in the Field of Deep Subsurface Processes*. Izd. AN SSSR, Moscow, 1962, pp. 51–56 (in Russian).

161. Dobrynin, V.M.: *Deformations and Changes in the Physical Properties of Oil and Gas Reservoirs*. Nedra, Moscow, 1970 (in Russian).

162. Avchan, G.M., Matveenko, A.A., Stefankevich, Z.B.: *Petrophysics of Sedimentary Rocks at Depth*. Nedra, Moscow, 1979 (in Russian).

163. Shchukin, E.D., Rebinder, P.A.: On the Mechanism of Elastic Aftereffect in Low-Concentration Structured Suspensions of Bentonite. *Kolloidnyy Zhurnal* 33, No. 3 (1971), pp. 450–458 (in Russian).

164. Maslov, N.N., 1968. *The Principles of Soil Mechanics and Engineering Geology*. Vysshaya Shkola, Moscow (in Russian).

165. Deryagin, B.V.: *What Is Friction*. Izd. AN SSSR, Moscow, 1963 (in Russian).

166. Sokolov, V.N., Osipov, V.I.: Microstructural Changes in Clays under Plane Shearing. *Inzhenernaya Geologiya* No. 6 (1983), pp. 6–21 (in Russian).

167. Rebinder, P.A.:, 1979. *Selected Works*. Vol.2, Nauka, Moscow (in Russian).

168. Rebinder, P.A. (ed.): *Physicochemical Mechanics of Dispersed Structures*. Nauka, Moscow, 1966 (in Russian).

169. Shchukin, E.D., Pertsev, N.V., Osipov, V.I., Zlochevskaya, R.I. (eds.): *Physicochemical Mechanics of Natural Dispersed Structures*, Izd. Mosk. Univ., Moscow, 1985 (in Russian).

170. Shchukin, E.D. (ed.): *Progress in Colloid Chemistry and Physicochemical Mechanics*, Nauka, Moscow, 1992 (in Russian).

171. Efremov, I.F., Us'yarov, O.G.: Interaction of Dispersed Particles at a Great Distance. Mutual Fixation of Dispersed Particles of Different Sizes and Forms. *Kolloidnyy Zhurnal* 34, No. 2 (1972), pp. 213–218 (in Russian).

172. Nesterov, I.I.: Compaction of clay rocks. *Sovetskaya Geologiya* No. 12 (1965), pp. 69–80 (in Russian).

173. Ushatinskiy, I.N., Nesterov, I.I., Zininburg, P.Ya.: The Screening Properties and the Classes of Shale Seals for Oil and Gas-Bearing Horizons of West Siberia. In: I.I. Nesterov (ed.): *Methods and Results of Studying Clays in West Siberia. Trudy Zap.-Sib. NIGNI* No. 35, Tyumen, 1970, pp. 98–122 (in Russian).

174. Erba, E., Premoli Silva, I., 1994. Orbitally Driven Cycles in Trace-Fossil Distribution from the Piobico Core (Late Albian, Centrale Hale). In: A.G. Plint (ed.): *Special Publications of the International Association of Sedimentology*, Blackwell Publ., New Mexico, pp. 221–255.

175. Allen, P.A. *Earth Surface Processes*. Blackwell Science, Oxford , 1997.

176. Bengtsson, H., Stevens, R.J.: Source and Grain-Size Influences upon the Clay Mineral Distribution in the Skagerrak and Northern Kattegan. *J. Clays and Clay Minerals* 33, No. 1 (1998), pp. 3–13.

177. Wells, J.T., Scholtz , C.A., Soreqhan, M.J.: Process of Sedimentation on a Lacustrine Border-Fault Margin: Interpretation of Cores from Lake Mala Wi, East Africa. *Sed. Res.* 69 (1999), pp. 816–831.

178. Strakhov, N.M.: On the Periodicity and Irreversible Evolution of Sediment Formation in the Earth's History. *Izvestiya AN SSSR, Ser. Geol.*, No. 6 (1949), pp. 20–111 (in Russian).

179. Timofeev, P.P.: *Jurassic Coal-Bearing Formation of South Siberia and the Conditions of Its Accumulation*. Trudy GIN AN SSSR No. 198, Nauka, Moscow, 1970 (in Russian).

180. Uspenskaya, N.Yu.: *Some Regularities in Oil and Gas Accumulation on Platforms*. Gostoptekhizdat, Moscow, (1952) (in Russian).

181. Kazarinov, V.P.: Stages of Oil and Gas Occurrence in the Mesozoic Sediments of Siberia. In: Y.A. Kosygin (ed.): *The Problems of Siberian Oil*. Izd. SO AN SSSR, Novosibirsk, 1963, pp. 46–51 (in Russian).

182. Angevine, C.I., Heller, P.L., Paola, C.: *Quantitative Sedimentary Basin Modeling*. AAPG, Tulsa, Oklakhoma, 1990.

183. Rateev, M.A.: *Distribution and Genesis of Clay Minerals in the Present and Ancient Sea Basins*. Trudy GIN AN SSSR No. 112, 1964 (in Russian).

184. Chen, Y., Chen, W., Park, A., Mu, J., Ortoleva, P., Cirf, A.: *A General Coupled Reaction – Transport Model and Simulator*. Indiana University, Indianapolis, 1994.

185. Wallbridge, S., Voulqaris, G., Tomlinson, B.N. and Collins, M.B.: Initial Motion and Pivoting Characteristics of Sand Particles in Uniform and Heterogeneous Beds: Experiments and Modeling. *Sedimentology* 46 (1999), pp. 17–32.

186. Ginzburg, I.I.: Problems in the Reaction Energetics of the Weathering Processes of some Aluminosilicates. In: I.I. Ginzburg (ed.): *The Weathering Crust*. No. 5. Izd. AN SSSR, Moscow, 1963, pp. 25–119 (in Russian).

187. Lisitsyna, N.A.: *Removal of Chemical Elements in the Weathering of Basic Rocks*. Trudy GIN No. 231, 1973 (in Russian).

188. Gradusov, B.P.: Location of Clay Material Profiles in Soils. *Doklady AN SSSR* 209, No. 5 (1973), pp. 1164–1167 (in Russian).

189. Golke, M.: *Patterns of Stress in Sedimentary Basins and the Dynamics of Pull Apart Basin Formation*. Ph.D. Thesis. Free University. Amsterdam, 1996.

190. Young, M.J., Gawthorpe, R.L., Sharp, J.R.: Sedimentology and Sequence Stratigraphy of a Transfer Zone Coarse-Grained Delta, Miocene Suez Rift, Egypt. *Sedimentology* 47 (2000), pp. 1081–1104.

191. Prozorovich, G.Eh.: 1967. The Principal Properties of Clay Beds and Clays as Impermeable Screens for Oil and Gas. In: F.G. Gurari, A.E. Kontorovich (eds.): *Geology and Oil and Gas Occurrence in the South-East of the West-Siberian Plate. Tr. SNIIGGiMS,* Iss, 65. Izd. SNIIGGiMS, Novosibirsk, pp. 110–124 (in Russian).

192. Timofeev, P.P., Eremeev, V.V., Bogolyubova, L.I.: 1985. Evolution of the Atlantic Ocean in the Mesozoic and the Problem of Black Shales. In: P.P. Timofeev (ed.): *Organic Matter.* Nauka, M., pp. 179–192 (in Russian).

193. Kearey, Ph.: *The Encyclopedia of the Solid Earth Sciences*. Blackwell Science, Oxford, 1995.

194. Hill, P.R., Lewis, C.P., Desmarais, S., Kauppaymuthoo, V., Rais, H.: The Mackenzie Delta: Sedimentary Processes and Facies of a High-Latitude, Fine-Grained Delta. *Sedimentology* 48 (2001), pp. 1047–1078.

195. Nesterov, I.I., Ushatinskiy, I.N.: The Screening Properties of Clay Rocks over Oil and Gas Pools of West Siberia. *Sovetskaya Geologiya* No. 5 (1971), pp. 51–63 (in Russian).

196. Filippov, B.P.: *Impermeability of Covers over Reservoir Rocks as a Factor of Oil and Gas Field Formation*. VNIGRI, Leningrad, 1964 (in Russian).

197. Antonova, T.F.: On the Classification of Clay Seals within the Geological Section of the Central West-Siberian Lowland. *Trudy SNIGGIMS, Novosibirsk* 47 (1966), pp. 128–131.

198. Zhemchuzhnikov, Yu.A.: The Cyclic Character of the Coal-Bearing Sequence Structure, the Periodicity of Sediment Accumulation, and the Methods for Their Study. *Trudy IGN AN SSSR, Ugol'naya Ser.* 90, No. 2 (1947). pp. 7–18 (in Russian).

199. Timofeev, P.P.: *Geology and Facies of the Jurassic Coal-Bearing Formation of South Siberia*. Trudy GIN AN SSSR No. 197, Nauka, Moscow, 1969 (in Russian).

200. Nesterov, I.I.: *Criteria for the Petroleum Occurrence Prediction*. Trudy. Zap.-Sib. NIGRI No. 15, Nedra, Moscow, 1969 (in Russian).

201. Kaplan, M.E., Vereninova, T.A., Galerkina, S.G.: *Peculiarities of the Change in Clay Sediment Properties in the Novyy Port Field (South Yamal)*. Trudy VNIGRI No. 326, 1973, pp. 188–192 (in Russian).

202. Lebedeva, G.V.: *Lithological Peculiarities of the Kynov Horizon in the Pechora Depression in Relation to Their Screening Properties*. Trudy VNIGRI No. 326, 1973, pp. 104–116 (in Russian).

203. Kheirov, M.B.: The Influence of Sedimentary Rock Depths on the Transformation of Clay Minerals. *Izvestiya AN SSSR, Ser. Geol.* No. 8 (1979), pp. 144–151 (in Russian).

204. Kaplan, M.E.: Reservoir Properties of the Mesozoic Terrigenous Complex in the Bureya Trough. *Trudy VNIGRI* No. 326 (1973), pp. 98–103 (in Russian).

205. Bogolyubova, L.I.: *Organic Matter of Recent and Fossil Sediments.* Dr. Sc. (Geol.-Min.) Thesis (Author's abstract). GIN AN Russia, Moscow, 1992 (in Russian).

206. Grim, R.E.: *Clay Mineralogy.* New York, McGraw-Hill, 1953.

207. Konyukhov, A.I.: *Sedimentary Formations in the Continent-to-Ocean Transition Zones.* Nauka, Moscow, 1987 (in Russian).

208. Kotel'nikov, D.D., Konyukhov, A.I.: *Clay Minerals of Sedimentary Rocks.* Nedra, Moscow, 1986 (in Russian).

209. Rateev, M.A., Pokidin, A.K., Kheirov, M.B.: Clay Minerals, Their Distribution and Genesis in the Generalized Stratigraphic Section of Ayatolly-Marine. In: N.M. Strakhov (ed.): *Postsedimentary Alterations of Quaternary and Pliocene Sediments of the Baku Archipelago.* Nauka, Moscow, 1965, pp. 89–112 (in Russian).

210. Ushatinskiy, I.N.: Peculiarities of Clay Material Mineralogy in Productive Sediments of the West-Siberian Lowland. *Neft'i gaz Tyumeni* No. 2 (1969), pp. 5–11 (in Russian).

211. Shabaeva, E.A., Chulkova, V.: Mineralogy and Physicochemical Features of Clay Seals over the Mesozoic Productive Horizons of the Bukhara-Khiva Region. *Geologiya Nefti i Gaza* No. 9 (1966), pp. 47–52 (in Russian).

212. Van der Meer, J.J.M., and Warren, W.P.: Sedimentology of Late Glacial Clays in Lacustine Basins, Central Ireland. *Quatern. Sci. Rev.* 16 (1997), pp. 779–791.

213. Nadean, P.H.: An Experimental Study of the Effect of Diagenetic Clay Minerals on Reservoir Sands. *Clay and Clay Minerals* 46. No. 1 (1998), pp. 18–26.

214. Abboft, S.T.: Detached Mud Prism Origin of High Stand Systems Tracts from Mid-Pleistocene Sequences, Wanganul Basin, New Zealand. *Sedimentology* 47 (2000), pp. 15–30.

215. Bachu, S., Cuthiell, D., Kramers, J.: Effects of Core Scale Heterogeneities on Fluid Flow in a Clastic Reservoir. In: I.M. Stivenson (ed.): *Pap. Geol. Surv. Can.,* Academic Press, Ottawa, 1990, pp. 191–209.

216. Rosett, D.F.: Soft-Sediment Deformation Structures in Late Albian to Cenomanian Deposits, Sao Luis Basin, Northern Brazil: Evidence for Paleoseismicity. *Sedimentology* 46 (1999), pp. 1065–1081.

217. Tarasevich, Yu.I., Ovcharenko, F.D.: The Nature of Water Interaction with the Surface of Montmorillonite. *Ukr. Khim. Zhurnal* 33, No. 5 (1967), pp. 505–512 (in Russian).

218. Zlochevskaya, R.I., Sergeev, E.M.: General Concepts of the Process of Clay Soil Hydration. In: E.M. Sergeev (ed.): *Problems in the Engineering*

Geology and Soil Science No. 2. Izd. Mosk. Univ., Moscow, 1968, pp. 85–94 (in Russian).

219. Katsube, T.J.: Statistical Analysis of Pore-Size Distribution Data for Tight Shales from the Scotland Shelf. In: D.W. Gibson (ed.): *Pap. Geol. Surv. Can.*, Academic Press, Ottawa, 1993, pp. 365–372.

220. Drits, V.A., Lindgreen, H., Salyu, A.L.: Determination of the Content and Distribution of Fixed Ammonium in Illite by X-Ray Diffraction. Application of North Sea Illite-Smectite. *Amer. Mineralogist* 82 (1997), pp. 79–87.

221. Zuther, M., Brockump, O., Glaner, N.: Composition and Origin of Clay Minerals in Holocene Sediments from the South-Western North Sea. *Sedimentology* 47 (2000), p. 119–134.

222. Push, R., Arnold, M. The Sensitivity of Artificially Sedimented Organic-free Illitic Clay. *Engng. Geol.* 3, pp. 135–148.

223. Nichiporovich, S.P., Khil'ko, V.V.: The Influence of the Crystalline Structure Peculiarities of Clay Minerals on the Processes of Structure Formation in Their Dispersions. In: P.A. Rebinder (ed.): *Physicochemical Mechanics of Dispersed Structures*. Nauka, Moscow, 1966, pp. 141–145 (in Russian).

224. Lipatov, S.M.: *Physicochemistry of Colloids*. Goskhimizdat, Moscow-Leningrad, 1948 (in Russian).

225. Kruglitskiy, N.N.: *Physicochemical Principles of Regulating the Properties of Clay Mineral Dispersions*. Naukova Dumka, Kiev, 1968 (in Russian).

226. Alison, P., Jones, P., Omoto, K.: Towards Establishing Criteria for Identifying Trigger Mechanisms for Soft-Sediment Deformation: a Case Study of Late Pleistocene Lacustrine Sands and Clays, Onikobe and Nakayamadaira Basins, Northeastern Japan. *Sedimentology* 47 (2000), pp. 1211–1226.

227. Pollastro, R.M.: Clay Minerals as Geothermometers Indicators of Thermalmaturity for Hydrocarbon Exploration. *Bull. U.S. Geol. Survey* No. 2007 (1992), pp. 61–65.

228. Bjorkum, P.A., Walderhaus, O., Aase, N.E.: A Model for the Effect of Illitization on Porosity and Quartz Cementation on Sand Stones. *Sedimentary Petrology* 63, No. 6 (1993), pp. 1089–1091.

229. Monyushko, A.M.: On the Influence of Tectonic Factor on the Formation of the Engineering–Geological Properties of Clay Rocks. *Inzhenernaya Geologiya* No. 3 (1979), pp. 55–64 (in Russian).

230. Vorob'ev, V.N.: On the Influence of Dictating Layers on the Intensity of Rock Fracturing. *Neftegazovaya Geologiya i Geofizika* No. 10 (1969), pp. 20–22 (in Russian).

231. Rebinder, P.A., Shchukin, E.D., Margolis, L.Ya.: On the Mechanical Strength of Porous Dispersed Bodies. *Doklady AN SSSR* 154, No. 3 (1964), pp. 695–698 (in Russian).

232. Shchukin, E.D.: Some Problems in the Physicochemical Theory of the Strength of Dispersed Structures. In: F.D. Ovcharenko (ed.): *Physicochemical Mechanics and Lyophility of Dispersed Systems* 13 (1981), pp. 46–53 (in Russian).

233. Sokolov, V.N.: Physicochemical Aspects of the Mechanical Behavior of Clay Soils. *Inzhenernaya Geologiya,* No. 4 (1985), pp. 28–41 (in Russian).

234. Amelina, E.A., Shchukin, E.D.: Study of Some Regularities in the Formation of Contacts in Porous Dispersed Structures. *Kolloidnyy Zhurnal* 32, No. 6 (1970), pp. 795–800 (in Russian).

235. Babak, V.G.: *The Strength of Porous Solids*. Candidate Sci. Thesis (Author's abstract). Moscow State Univ., Moscow, 1974 (in Russian).

236. Field, W.G., Towards the Statistical Definition of a Granular Mass. In: J. Collins (ed.): *Proc. 4th Australia–New Zealand Conf. on Solid Mechanics*. New Zeland Geomechanics Society, Auckland, 1963., pp. 143–148.

237. Gray, W.A.: *The Packing of Solid Particles*. Chapman and Hall, London, 1968.

238. Sokolov, V.N.: Models of Clay Soil Microstructures. *Inzhenernaya Geologiya* No. 6, (1991), pp. 32–40 (in Russian).

239. Pashley, R.M., Israelachvili, J.N.: A Composition of Surface Force and Interfacial Properties of Mica in Purified Surfactant Solution. *Colloid and Surfaces* 2 (1981), pp. 169–187.

240. Butt, H.-J.: A Technique for Measuring the Force between a Colloidal Particle in Water and a Bubble. *J. Colloid Interface Sci.* 166 (1994), pp. 109–117.

241. Ducker, W.A., Xu, Z., Israelachvili, J.N.: Measurement of Hydrophobic and DLVO Forces in Bubble–Surface Interactions in Aqueous Solutions. *Langmuir* 10 (1994), pp. 3279–3289.

242. Preuss, M., Butt, H.-J.: Direct Measurement of Particle–Bubble Interactions in Aqueous Electrolyte: Dependence on Surfactant. *Langmuir* 14 (1998), pp. 3164–3174.

243. Preuss, M.: Direct Measurement of Forces between Particles and Bubbles. *Int. J. Miner. Process* 56 (1999), pp. 99–115.

244. Tyrrell, J.W. G., Attard, P.: Images of Nanobubbles on Hydrophobic Surfaces and Their Interactions. *Phys. Rev. Lett.* 87, 176104 (2001).

245. Parker, J.L., Claesson, P.M., Attard, P.: Bubbles, Cavities, and the Long-Ranged Attraction between Hydrophobic Surfaces. *J. Phys. Chem.* 98, pp. 8468–8480 (1994).

246. Carambassis, A., Jonker, L.C., Attard, P., and Rutland, M.W.: Forces Measured between Hydrophobic Surfaces due to a Sub-microscopic Bridging Bubble. *Phys. Rev. Lett.* 80, pp. 5357–5360 (1998).

247. Mahnke, J., Stearnes, J., Hayes, R.A., Fornasiero, D., and Ralston, J.: The Influence of Dissolved Gas on the Interactions between Surfaces of Different Hydrophobicity in Aqueous Media. Part I. Measurement of Interaction Forces. *Physical Chemistry Chemical Physics*, 1, 2793–2798, (1999).

248. Vaganov, V.P.: *Experimental Study of Physicochemical Regularities in the Formation of Crystallization Contacts in the Intergrowing of Individual Crystals*. Candidate Sci. Thesis (Author's abstract). Moscow State Univ., Moscow, 1975 (in Russian).

249. Babak, V.G., Kozub, S.P., Sokolov, V.N., Osipov, V.I.: Methods for the Precision Measurement of the Interaction Energy of Condensed Bodies Under Different Physicochemical Conditions. *Izvestiya AN SSSR, Ser. Fiz.* 41 (1977), pp. 2401–240 (in Russian).

250. Sokolov, V.N., Yurkovets, D.I., Ragulina, O.V., Mel'nik, V.N.: Computer-Controlled System for the Study of Micromorphology of the Surface of Solids by SEM Images. *Surface Investigation* 14 (1998), pp. 33–41.

251. Street, N., Buchanan, A.S.: The ξ-Potential of Kaolinite Particles. *Austral. J. Chem.* 9, No. 4 (1966), pp. 450–466.

252. Van Olphen, H.: *An Introduction to Clay Colloid Chemistry (2nd. Ed.)* A. Willey-Interscience Publ., New York, London, 1977.

253. Hurst, C.A., Jordine, E.St.: Role of Electrostatic Energy Barriers in the Expansion of Lamillar Crystals. *J. Chem. Physics* 41, No. 9 (1964), pp. 23–27.

254. Arulanandan, K., Scott, S.S., Spiegler, K.S.: Soil Structure Evaluation by the Use of Radio Frequency Electrical Dispersion. In: L. Barden, R. Push (eds.): *Proc. Int. Symp. Soil Structure. Gothenburg.* Swedish Geotechnical Society, Stockholm, 1973, pp. 29–49.

255. Popov, I.V. (ed.): *Engineering-Geological Properties of Rocks and the Problems in Lithogenesis*, Nauka, Moscow, 1965 (in Russian).

256. Tuezova, N.A., Dorognitskaya, L.M., Demina, R.G., Bryuzgina, N.I. (eds.): *Physical Properties of Rocks in the West-Siberian Oil- and Gas-Bearing Province.* Nedra, Moscow, 1975 (in Russian).

257. Popov, I.V.: *Engineering Geology of the USSR. Part 1.* Izd. Mosk. Univ., Moscow, 1961 (in Russian).

258. Protod'yakonov, M.M., Teder, R.I., Il'inskaya, E.I., Yakobashvili, O.P., et al. *Distribution and Correlation of Indices of Rocks' Physical Properties. A handbook.* Nedra, Moscow, 1981 (in Russian).

259. Dudushkina, K.I., Bobrov, G.F.: *Deformation Properties of Rocks of Deep Horizons.* Nedra, Moscow, 1974 (in Russian).

260. Beron, A.I. (ed.): *Rock Properties under Stresses of Different Kinds and Regimes.* Nedra, Moscow, 1984 (in Russian).

Subject Index

absolute permeability to gas 133, 134
acids
  amino 4, 40
  fulvic 40, 41
  humic 4, 12, 13, 40, 41
  silicic 61
adsorptional reduction of the strength 117
adsorption centers 5, 47, 171
AFM cantilever 206
aggradation of clay minerals 39
aggregates 9, 11–13, 17, 23–25, 32, 33, 40, 41,
    48, 51, 56, 70, 92, 113, 127, 131, 138, 148,
    155, 164, 167, 170, 171, 200
aggregation 4, 13, 17, 21–27, 31, 36, 152, 157
aging of clay sediments 41
albumen 4, 40
alluvial valley 127
alpine geosynclinal 121
alumo-silica gels 39
ammonia 4
analysis
  chemical 61
  granulometric 11, 226, 228, 239, 249
  lithological-facies 126
  mineral 239
  orientation 35
  quantitative microstructural 34, 211, 212, 249
  X-ray 61
angle between contacting particles (θ) 216, 226
anomalous horizons in a clay sequence 87
aqueous medium 17, 27, 61, 155, 206
argillite 13, 53, 54, 70, 80, 95–97, 100,
    103–105, 118, 143, 203, 205, 212, 213,
    221, 231, 241, 244, 245
  bituminous silty 143
argillization of rock 87, 184
arid epoch 123
aride zone 13, 48, 49
authigenous dolomites 138, 139
average force of cohesion between particles (P_1)
    197

basin
  accumulation 127
  Bureya 140, 144–146, 173, 176
  Caspian 144, 145
  drainage 127
  marine sedimentation 125
  oil- and gas-bearing 76, 121, 122, 126, 127,
    130–132, 137, 172, 186
  Peri-Caspian 173
  sedimentary 3, 67, 70, 76, 110, 113, 121–123,
    125–131, 137, 139, 140, 142, 143, 147, 148,
    150, 154, 158, 164, 172–174, 176, 178
  South-Yamal 176
  South-Caspian 48
  Timan-Pechora 139, 143
  Volga-Ural 139, 144
  West-Sibirian 121, 122, 132, 139, 141–143,
    173, 174, 176
  Yamal 139, 143
  Zeya-Bureya 176
bedrock 11, 123
behaviour
  of clay rocks 96
  deformation 101–105, 108, 112
  rheological 109, 111
bitumen 40, 41, 141
black shales 130
bonds
  ionic-electrostatic 56, 115
  structural (see structural bonds)
boundary friction 59, 106
brittleness 97, 117, 169, 172, 183, 185, 192, 231
brown coal 67

calcite 3, 13
capacity
  absorption 131, 132, 147, 152, 155, 158
  exchange 92, 148, 150, 151, 154, 157, 158,
    160, 161
  filtering 189
  sorption 148

capillaries 7
carbonates 11, 13, 38, 133, 134
catagenesis 37, 45–48, 50, 51, 55, 59, 63–65,
    67–69, 72–74, 79–82, 85–89, 95–97, 100,
    102, 103, 105–110, 112, 115–118, 154,
    165, 168, 171, 172, 174, 175, 178, 179,
    180, 182, 184, 185, 187, 188–192, 211,
    212, 214, 217, 220, 221, 225, 230
catagenetic transformations of rocks 46
cation "bridges" 217
cations
    calcium 131, 152, 157
    exchange 6, 56, 66, 91, 94, 134, 148, 151,
        163, 191, 217
    higher-valence 92
    interlayer 87
    magnesium 151, 152, 157
    multivalent 93, 165
    potassium 40, 132, 206
    sodium 131, 151
    univalent 92
cations-compensators 4
chemical composition of filtering fluid 92, 188
chlorite 3, 48, 50, 68–70, 74, 123, 125–128, 130,
    134, 135, 139, 147, 150, 152, 155, 163, 166,
    169, 170, 179, 180, 182, 216, 226, 241
classification of clay seals 128, 131, 132
clastic grains 16, 46, 113, 116, 169–171, 187,
    191
clastic material 139, 140, 168, 170
clay coats 44, 226
clay dispersion 17, 23, 24, 27, 28, 30, 33
clay envelops 171
clay minerals 3, 4, 5, 7, 8, 9, 11, 13, 38, 39, 40,
    41, 48, 49, 50, 61, 64, 66, 67, 69,70, 75,
    84, 92, 113, 123, 125, 126, 129, 131–135,
    147, 149, 156, 169, 170, 185, 212
clay particles 4, 6, 9, 10, 11, 13, 22, 27, 28,
    29, 30, 31, 40, 42, 44, 46, 51, 53, 55, 56,
    57, 58, 59, 60, 80, 85, 92, 94, 102, 116,
    118, 127, 131, 150, 152, 155, 164, 165,
    170, 171, 202, 203, 204, 209, 211, 212,
    214–218, 221, 225–228, 235, 236,
    248–250
    primary 9, 11
clay rocks 8, 9, 13, 14, 17, 46–48, 50, 51,
    53–56, 58, 61, 63–66, 70, 71, 73–75, 81,
    84, 86, 87, 89–91, 94–98, 101, 102,
    104–109, 112, 117, 118, 151, 163, 172,
    180, 181, 183, 185, 187, 188, 190, 197,
    200, 202–204, 211, 212, 214, 215, 217,
    220–222, 226, 228, 229, 231, 232, 239, 255
clay rock sequences 74
clay schists 203, 204, 212, 221, 231
clay seals 1, 7–9, 49, 71, 87, 95–97, 100, 118,
    119, 121, 122, 125, 128, 131–133, 135,
    137, 138, 140–144, 147–149, 156,
    158–161, 163, 173–175, 180–182, 185,
    188, 190–193, 195, 197, 198, 203, 204,
    211–216, 219, 220–223, 225–236, 239,
    240, 242, 244, 246–257
    regionally persistent 122
clay sediments 3, 5, 7, 13, 27, 28, 30–38, 41, 42,
    45, 46, 64, 67, 71–74, 76–78, 81–85, 88,
    95, 99, 101–104, 107–109, 111, 115, 118,
    130–134, 138, 139–142, 145, 147, 151, 152,
    154, 155, 163, 165–167, 169, 171–182,
    185–187, 189, 190, 193, 216, 220, 230, 231
    deep water 147, 163, 178, 180, 192
    relatively deep water 178, 180, 185, 192
    shallow water 46, 67, 85, 169, 171, 179, 180
clay sequences 8, 46, 47, 117, 121, 126, 132,
    133, 135, 181, 188, 191, 193
    Jurassic 132
    Lower-Cretaceous 132
clays 7–9, 11–17, 26, 30, 38, 42, 44, 48, 49, 53,
    61, 66, 69, 73, 76, 81, 84, 89, 91–94, 96,
    98–100, 103, 105–110, 112, 122, 123, 126,
    130, 131, 133, 134, 138–146, 148, 150,
    157, 171–173, 177, 178, 180–183, 185,
    188–193, 201, 203–205, 212, 219, 221,
    226, 231
    chlorite 123
    fine grained 131, 135, 139, 140, 141, 143,
        145, 147
    hydromica 94, 123
    kaolinite 92, 93, 95, 123
    low-silt 139, 140, 142, 145
    montmorillonite 11, 91, 93–95, 123
    silty 130, 132, 134, 138, 141, 143, 144, 146,
        204, 221, 226
claystone 98
clearance 20, 21, 30, 57, 165, 171, 205–207,
    216, 222, 223
cleavage 46, 64, 81, 118
cleavage planes 178, 184
climate 123–125
    humid 123, 125
coagulation 4, 17, 22–27, 29, 32, 58, 68, 92,
    105, 113, 167, 169

coal metamorphism 67
coals 50, 69
    gas fat 50, 69
    long-flame 50, 69
coccolithophorids 3, 13, 33
coefficient
    correlation (R) 148, 220, 221, 225, 226, 229
    diffusion 129, 133–135
    of filtration (K) 90
    of internal friction (tg φ) 104
    of tortuosity (ε) 90
    of uniaxial compressibility 98, 99, 100
cohesion 27–30, 40, 45, 56, 57, 59, 83,
        104–107, 114–116, 118, 197, 198,
        205–210, 216, 223
    of a reversible character at water content W
        and density ρ (C_pw) 105
    of irreversible character (C_ε) 105
colloid aging 113
colloid stability 5
colloidal activity 41
colloidal hydroxides 4
colloids 4, 5, 40, 46
compactibility 79
compaction
    gravitational 41, 42, 45, 81, 168, 176, 178,
        179, 191
    rock 46, 47, 56, 66, 74, 109, 115, 118, 165,
        202, 212
compaction stages 77
complex
    absorbed 129, 130, 131, 132, 135, 137, 147,
        152, 158, 163, 166
    deep-water fauna 130
    exchange 39, 61, 92, 93, 127, 133, 151, 155,
        157, 164, 165, 167, 169, 172, 182, 185, 191
    organic-mineral 13, 40, 61
composition
    chemical 91, 92, 93, 123, 188
    grain-size 130
    granulometric 137, 148, 155, 158, 215, 221,
        222, 228
    mineral 9, 11, 26, 33, 38, 48, 74, 86, 91, 106,
        107, 121, 131–133, 147–150, 152, 154,
        156, 163, 166, 169, 172, 173, 175, 180,
        188, 191, 207, 212, 215, 226
    of fauna 137
    of the absorbed complex 137
    polymineral 33, 134
compressibility 97–100, 107, 114, 165

concentration 18, 24–26, 31, 32, 41, 46, 70, 82,
        85, 87, 92, 93, 112, 113, 164, 165, 167,
        178, 182, 197, 205, 209, 236
    electrolyte 25, 209
    pore solution 46, 87, 92, 93
    salt 93, 165, 167, 209
    solid phase 167
conditions
    aerobic 4, 40
    anaerobic 4, 40
    climatic 127
    facies-geochemical 131, 151
    geochemical 60, 83, 172
    paleogeographic 126
    physicochemical 46, 66, 207
    physiogeographic 127
    pressure-temperature 46, 66, 71, 89
    tectonic 127
    thermodynamic 91
consistence 83, 84, 88, 89, 113, 114, 215
    fluid 67, 89
    fluid-plastic 83, 89, 114
    latent-fluid 68, 83, 89, 113
    plastic 68, 89
    semi-solid 68, 69, 83, 89, 108, 115, 215
    soft-plastic 83, 89, 114
    solid 68, 83, 89
    tight-plastic 83, 89, 108, 115, 215
consolidated sedimentary rock 37
contact area 59, 60, 63, 118, 197, 201, 207,
        216, 218, 221, 223, 225
contacts
    close coagulation 28, 36, 45, 67, 68, 79, 89,
        102, 108, 111, 113, 169, 205, 212, 214,
        217, 220, 221, 223, 225, 230, 231
    coagulation 27, 28, 30, 68, 79, 81, 82,
        101–103, 105–107, 111–115, 165, 182,
        197, 205, 219–221
    distant coagulation 36, 45, 82, 89, 99, 101,
        111, 113, 114, 168, 211, 212, 220, 221
    phase cementation 62, 68, 82, 89, 109
    phase 26, 57–60, 62, 63, 68, 80, 82, 89, 97,
        105, 106, 109, 112, 116, 168, 178, 180,
        182, 185, 205, 206, 207, 212, 214, 217,
        219, 220, 221, 223, 225, 226, 230, 231
    phase crystallization 60, 68, 82, 89, 109
    transitional 26, 56–59, 68, 80, 82, 89, 102,
        103, 105, 106, 109, 112, 115, 116, 168,
        178, 180, 185, 205, 206, 207, 212, 214,
        217, 219–221, 223, 225, 226, 230, 231

coordination number (z) 199, 201, 227, 237
correlation equation (dependence) 150, 152,
    155–157, 186
Coulomb-Moor's theory 107, 114
creep 82, 86, 110–114, 116, 176, 178, 180, 183,
    185, 191
creep limit 110–112
crossed plates of mica 207
crystal lattice 5, 7, 40, 92, 135
    expanding 92
    rigid 92
crystallization 8, 40, 58–60, 70, 80, 87, 116,
    118, 185, 221, 225
    of amorphous substances 46
    of colloids 46
crystals of dehydrous gypsum 207
cuffs inside of pores 85
current 206, 207
currents
    near-shore 127, 155, 169
    turbidity 128, 154
curves of pore distribution according to the
    equivalent diameters 35

Darcy (D) 90
Darcy's law 94
Debye's radius of ionic atmosphere 6
deep Aralsor well 74
deep-water drilling 46
deep-water shelf and adjacent sea 125, 147,
    159, 163, 185, 188, 193
defects of the rock structure 109
deformation curve 99–102, 107–109, 115
deformation 26, 45, 58, 60, 66, 77, 87–89, 97,
    101–105, 107–109, 111–117, 165, 169,
    175, 176, 178, 182, 183, 185, 191, 207, 221
    lateral 98
    vertical 98
degree of lithification 66, 101, 109, 185, 231,
    239
degree or index of orientation (K_a) 16
density 35, 42, 44, 52, 53, 69, 71–75, 84, 90,
    99, 102, 105, 106, 117, 132, 171, 172, 201,
    217, 218, 223, 227, 236, 251
    clay 69, 73, 102, 132, 172
    deposit 72
    dry rock (ρ_d) 71, 73, 74, 84
    mineral part (ρ_m) 71, 72, 74
    rock (ρ) 71, 73, 74, 117
    sediment 72, 99, 106, 171

surface charge 218
depth
    of burial 66, 154, 155, 177, 183
    of occurrence 50, 78, 149, 220, 223, 226, 227,
        232, 233, 239, 240, 242, 246–253, 255, 256
    of sediment burial 76, 77, 154, 155
destruction
    avalanche-like 111, 113
    brittle 87, 103, 109, 112, 115, 116, 169, 183
    quick (liquefaction) 111
    semi-brittle 103, 112, 115
    viscous 112
detritus 3, 123, 141
device 205–208
dewatering of sediments 95
diagenesis 37, 38, 40–42, 45, 48, 53, 64–69, 73,
    74, 79, 81, 82, 85, 88, 89, 95, 98, 99, 101,
    102, 108–115, 165, 168, 171, 178, 180, 182,
    188, 192, 211, 212, 216, 217, 220, 221, 225
diatoms 3, 13
dickite 64
dielectric permittivity 6, 21, 217, 218
diffusion 40, 60, 85, 86, 129, 132, 165, 168
dioctahedral hydromica 2M$_1$ 64
direct measurement
    of cohesion forces 205, 207
    of the strength of individual contacts 205
dislocations 109
dispersed clay material 171
dispersion 5, 8, 17, 23, 24, 27, 28, 30, 33, 92,
    112, 131, 138, 147, 148, 150, 152, 154,
    169, 182, 185, 191, 193
    aqueous 31, 165
DLVO theory 17, 22, 29, 205, 209
drainage ability 179
drainage horizon 168
draining interlayers 46, 47
dynamic effects 109, 113
dynamics
    active 128
    relatively active 128
    very active 128
    water 127–129, 134
dynamometer 205, 208

early diagenesis substage 38, 67, 95, 114, 165
early substage of catagenesis 46
earthquake 109
effective cross section 94
elastic decompaction of rocks 76

elastic membrane 94
elastic wave velocities 100
electric resistivity 100
electrokinetic phenomena 4
electrokinetic potential 4
electrolyte 17, 25, 26, 205, 207, 209
  alkaline 210
electrostatic component of the wedging pressure
  17, 18
electrostatic field 4, 17
energy 4, 5, 7–9, 17, 19–28, 30, 38, 45, 56, 57,
  60, 61, 66, 84, 87,88, 93, 94, 97, 109, 112,
  113, 165, 167, 183, 197, 205, 216
  free interphase (σ) 197
  interaction 5, 20–24, 26, 28, 205, 206, 216
  of molecular interaction 23, 205
  surface 4, 24, 61, 112, 113, 165, 167, 183
energy curve 23
energy states of water 88
environment
  deep-water 67, 140, 141, 142
  deep-water marine 122, 140, 141
  deep-water shelf 133, 138, 139, 141, 143
  depositional 8, 33, 38, 42, 46, 65, 121, 125,
    128, 131, 132, 135, 137–139, 145, 147,
    155, 163, 166, 172, 180, 181, 185, 193, 212
  geochemical 132
  medium-depth 130, 141
  middle-shelf 133, 140, 183
  middle-shelf and peripheral part of delta 135,
    166, 193
  natural 81
  near-shore 3
  shallow shelf 134
  shallow-water 130, 133–135, 138, 139,
    141, 143, 145, 155, 169–171, 187, 189,
    192, 193
  shelf 128, 133, 134, 138–141, 143, 145
epigenesis 37
epigenetic alterations 180
epochs of intense tectogenesis 125
equation
  power-function 226, 228, 233
  quadric polynomial 220, 225, 229, 231, 248,
    249, 253
equivalent diameter 35, 44, 45, 52–55, 79, 80
evolution geological 126
exchange
  ion 4, 165
  water 41, 123

exchangeable calcium 132
exchangeable sodium 132, 135, 148, 155
experimental database 239

fabric (see texture)
facies
  basinal 148
  coastal 139, 178
  of the deep-water shelf and adjacent sea basin
    163
  of deep-water shelf 126, 139, 147, 148, 159,
    163, 188
  deltaic 126
  marine basinal 139
  marine deep-water 141
  medium-depth shelf 126, 159, 176
  of clay and clay-silt sediments 138
  of coastal zone 155, 159, 169, 178, 179, 187,
    189
  of middle shelf 151, 166, 176, 177, 186, 188,
    190
  of peripheral part of delta 151, 159, 176,
    177, 186, 188, 190
  of shallow water shelf 126, 138, 139, 155,
    159, 169, 178, 189
  of the middle shelf and peripheral part of delta
    151, 166, 176, 177, 186, 188, 190, 193
  of the shallow-water shelf and coastal zone
    155, 159, 169, 178, 189
  shallow-water 179, 187, 193
  type 119, 137, 147, 158–161, 163, 172, 173,
    180, 185, 188, 190, 193
factors
  framework creep 178, 180, 185
  physicochemical 7, 31, 41, 172, 197, 198,
    205, 208, 209
  pressure–temperature 95, 148, 172, 185
  stabilizing 4, 24, 29, 31, 163
  structural-tectonic 181
features
  genetic 137, 138
  micromorphological 198
  morphometric 8, 9, 13, 16, 55, 64
  textural 137
feldspars 3, 38
ferrous compounds 3
ferrous kaolinite 124
ferrous-magnesium chlorite 125
filtrational anisotropy 46, 53
flanks of uplifts 183

flocculation 23, 24
foraminifers 3
force
    cohesion 30, 56, 197, 205–210, 216, 223
    of attraction 17, 19, 21–23, 26, 28, 41, 57,
        113, 216
    of contact interaction 205, 207
    of repulsion 17, 21, 23, 45, 210, 216
    pressing 58, 207, 209, 223–225
    chemical 19, 59, 60, 62
    interaction 5, 20–24, 26, 28, 29, 197, 205,
        206, 216
    ionic-electrostatic 29, 59, 60, 208, 210
    magnetic 29
    molecular 4, 14, 19–21, 23, 27, 28, 41, 67,
        113, 164, 170, 205, 216, 223
    osmotic 6, 84
force-measuring device 206
formation
    Alym 141
    Bazhenov 139
    Berezov 143
    Culomzin 140
    Gan'kin 143
    Georgiev 139
    Kuznetsov 141, 143
    Megion 140
    Pokur 141
    Tar 140
    Tyumen 139, 143
    Vasyugan 139
    Vartov 140
fraction
    clay (pelitic) (<0.005 mm) 3, 69, 130, 133,
        134, 148, 149, 153, 156, 159, 221, 240, 246
    fine grained 41
    sand (0.05–2 mm) 3, 132, 134, 135, 149, 153,
        156, 159
    silt (aleuritic) (0.05–0.005 mm) 3, 132, 134,
        135, 149, 153, 156, 159, 228, 240, 246
framework of the card house type 32
framework 5, 7, 9, 11, 22, 32, 113, 126, 168,
        169, 171, 172, 179
    structural 9, 168, 169, 171, 179
friction, or of shearing resistance angle (φ)
        104–108, 115, 116, 118

galvanometer pointer 206
general porosity-depth dependence 73, 173,
        177

genesis 42, 69, 121, 132, 137, 140, 144, 239
    alluvial 121
    continental 121, 140, 144, 145
    marine 121, 140, 144, 145
    near-shore 121, 144, 145
genetic types of pores 79
geodynamic regime 83
geological history 118, 123, 137
geometric features of the structure 9, 16, 55
geosynclines 46, 63
goniometric device of SEM 207
gradient
    filtration 94
    geothermal 46
    hydrodynamic 188
    initial 94
    pressure 14, 91, 94, 96
    temperature 94
grain size distribution 38, 40, 66, 71, 86, 137,
        154, 156, 159, 160, 191, 212
grains 9, 11–13, 16, 32, 33, 42–44, 46, 48, 51,
        58–60, 62, 63, 66, 71, 80, 113, 116, 128,
        129, 154, 156, 159, 166, 169–171, 179,
        180, 182, 185, 187, 189, 191, 192, 200,
        203, 204, 221–227, 251
gypsum 3, 74, 207

halloysite 10, 38
Hamaker's constant (A) 205, 222
helium-neon laser 208
heterocoagulation 24, 25, 41, 113, 170, 171
heterogeneity
    of microfracture 117
    of microstructure 182, 187, 191, 192
    lithologic 132
    structural 84, 87, 169, 171, 184
    structural bond 64, 87, 169, 178
heteropotential 5, 22, 28
heterovalent isomorphic substitutions 4
higher plants' tissues 139
highly-oriented clays and clay rocks 17
holder 207, 208
homogeneity
    of clay microstructure 69, 182, 183, 185, 191
    of sediments 168, 172, 177, 193
    of the rock 165, 191
humates 5, 40
humus-type organic matter 127, 134, 155
hydrocarbon fields 132
hydrocarbon traps 132

hydromica 3, 31, 38, 48, 50, 64, 68, 69, 73, 91, 96, 123, 169, 173, 216, 226
hydromicanization of montmorillonite 48, 87, 117, 184
hydrophobic particle interaction 206
hydrotroilite 38

illite 10, 68, 69, 125, 127–130, 133–135, 138, 139, 147, 148, 152, 155, 163, 166, 167, 169, 176–178, 182, 185, 188, 189, 192, 241
illitization 178, 185, 188, 191, 192
image analyzer 208
image processing system 212
index
    rock brittleness (fragility) 231
    vitrinite reflectance (R^a) 50, 66, 68, 69
interaction of microaggregates according to
    edge-to-edge 32, 167
    face-to-edge 25, 31, 32, 33, 44, 167, 170, 207
    face-to-edge at a low angle 31, 33, 51, 164
    face-to-face 9, 31, 33, 44, 53, 164; see also interaction of the basis–basis type
interaction of the basis–basis type 170
interference pictures 208
interlayer
    sand-silt 85, 125, 126, 134, 142
    swelling 87, 164, 173
interlayer interval 84, 87, 184
ionic-electrostatic bridges 57
iron hydroxides 5, 13, 74
iron smectites 11, 38
isomorphism 87
isotherm 47, 70, 87, 88, 117
isothermal boundary 86, 96, 183, 185

kaolinite 3, 9–11, 28, 31, 33, 38, 68, 69, 73, 91–94, 96, 125–128, 134, 135, 139, 147, 152, 155, 163, 166, 169–171, 179, 180, 182, 216, 226, 241

late diagenesis substage 45, 64, 67, 85, 95, 99, 101, 107, 108, 110–112, 114, 115, 165, 180, 216, 217, 220, 221, 225
late substage of catagenesis 69, 80, 82, 87, 88, 96, 165, 169
layer
    adsorbed cation 4
    diffuse 4, 17, 18, 25, 46, 56, 85, 92–94, 116, 169, 210, 216
    electrical double (EDL) 4, 17, 40, 87, 209, 216

swelling 33, 74, 125, 147, 152, 163, 166, 173, 174
length of clay particles (a) 215, 216, 235, 248
lenses of sand and silt material 169
lignin 4, 40
limestones 183
limit
    liquid (W_L) 36, 41, 67, 84, 85, 88, 89, 166
    plastic (W_P) 69, 84, 86, 89, 116
    swelling (W_S) 84
liquefaction 111
lithified clays 13, 30, 106, 109
lithogenesis stage 37, 38, 41, 46, 49, 64–68, 171, 180, 189, 216, 221
lithogenesis 37, 38, 48, 50, 61, 64–68, 71, 72, 74, 76, 78–84, 86–89, 91, 95–99, 104, 112, 117, 118, 132–134, 148–150, 152, 155, 163, 165–168, 172, 178–180, 182, 191, 193, 195, 197, 211, 212, 215
lithological-facies model 126
live cross section of pores 94, 96, 113, 117
loaded element 206
loam
    loess-like 200, 201
    marine 200, 202
low-temperature regional metamorphism 63

macrobodies 205
macropores 14
macroscopic strength of sample 211
magnetoelectric system 206, 207
    of galvanometer 206
main active window 234–236
manipulator 206
marine deep-water fauna 139
mechanical sedimentary differentiation 127, 129, 130, 147, 155
mechanical strength of clay seals 234, 257
medium-oriented clays and clay rocks 17
mercury method of porosity metering 75
metagenesis 37, 63, 70, 74, 81, 83, 212
metamorphism 37, 63, 67
method of multiple interference 205
mica plates 30, 205, 209, 210
micas 3, 38, 48, 56
microaggregates 9, 11–14, 16, 24, 32, 33, 36, 42, 44, 51, 53, 55, 79, 80, 82, 85, 86, 94, 113, 114, 118, 164–171, 176, 179, 180, 182, 188, 192, 200, 202–204
    sheet-like 202

microbalance with negative feedback 205
microblocks 13, 51
microcrystals 9, 13, 92, 207, 208, 223, 224
microfauna 9, 12, 13, 16, 33
microflora 9, 13, 16
microfracture formation 86, 87, 178, 182, 184, 192
microfractures 13, 76, 80, 81, 83, 87, 88, 96, 97, 104, 112, 116–118, 129, 133, 169, 172, 175, 178, 180, 182–192, 229, 255
microfracturing 8, 81, 97, 100, 133–135, 165, 169, 181–183, 185–189, 192, 232, 233
micropores 14, 16, 35, 42, 44, 45, 51–55, 75, 64, 79, 80, 81, 85–88, 95, 113–115, 117, 164, 167, 168, 171, 180, 188, 189, 191
microscope
    atomic force (AFM) 206
    long-focus 206
    scanning electron (SEM) 8, 11, 13, 44, 51, 53, 54, 79, 203, 207, 208, 212, 215, 226, 249
microstructure 8, 30–36, 42–45, 50–55, 64, 65, 69, 79, 108, 112, 113, 133, 163–168, 170–172, 174, 175, 180, 182, 183, 185, 188, 189, 191, 192, 197, 198, 200, 202, 204, 211–214, 222, 226
    blastic 64, 65
    honeycomb 31–36, 42, 45, 79, 113, 166, 167
    honeycomb-matrix 42
    laminar 51, 53, 54, 55
    large-honeycomb 113, 165–170
    large-pore 188, 189
    matrix 42–45, 53
    matrix-turbulent 55
    of a kaolinite sediment 170
    of a marine sediment composed of illite 167
    of a Na-montmorillonite sediment 164
    of clays 8, 171
    primary 30, 108, 112, 163, 191
    small honeycomb 85, 113, 164, 165
    turbulent 51–53, 55
    turbulent-laminar 55
middle substage of catagenesis 96
migration of hydrocarbons 41
minerals
    allothigenic 3, 38, 69
    authigenic 3, 38, 74, 81, 117, 132
    clay 3, 4, 5, 7, 8, 9, 11, 13, 38, 39, 40, 41, 48, 49, 50, 61, 64, 66, 67, 69,70, 75, 84, 92, 113, 123, 125, 126, 129, 131–135, 147, 149, 156, 169, 170, 185, 212

clay particles 4, 6, 9, 10, 11, 13, 22, 27, 28, 29, 30, 31, 40, 42, 44, 46, 51, 53, 55, 56, 57, 58, 59, 60, 80, 85, 92, 94, 102, 116, 118, 127, 131, 150, 152, 155, 164, 165, 170, 171, 202, 203, 204, 209, 211, 212, 214–218, 221, 225–228, 235, 236, 248–250
    heavy 137, 138
    light 137, 138, 139
    mixed-layered 3, 9, 10, 11, 33, 38, 48, 50, 68–70, 73, 74, 87, 125, 126, 130, 148, 149, 154, 155, 163–166, 168, 169, 173, 180, 185, 188, 189, 191, 192, 215, 241
    montmorillonite–hydromica 173
    non-swelling 125, 134, 135, 147, 149, 150, 152, 154, 156, 160, 165, 169, 185, 216
    rock-forming 3, 123, 187
    swelling 48, 85, 149, 152, 154, 159, 160, 165, 168, 212, 214–216
mixed-layer illite-montmorillonite varieties 130, 133–135, 138, 147
mixed-layer montmorillonite-illite varieties 138
mixed-layer phases 3, 48, 69, 149
model
    bi-dispersed globular 201, 204, 223, 227, 249, 251
    Bingham's 111
    globular 198–201
    of clay sediment microstructures 31
    Shvedov's 110
    twisting card house 202–204, 211, 225, 249, 250
modulus
    bulk (K) 98
    fracturing 186, 187
    total strain (secant modulus) (E_t) 100–103, 114–116, 118
    elastic strain (Young's modulus) (E) 58, 100, 101, 103, 104, 115, 116, 118
molecular attraction 19, 20, 27, 28, 210, 222
molecular component of the wedging pressure 19
mollusk 148
montmorillonite 3, 9–11, 31, 33, 38, 47, 48, 68, 69, 73, 74, 85, 87, 91–93, 125, 126, 128, 130, 133, 138, 147, 149, 150, 152, 154, 163–166, 173, 174, 184, 191, 218, 241
morphometric features of the structure 9, 16
mudstones 63

naphthalene 207
neocrystallization 42, 74, 81, 82, 88, 117
non-layered silicate recrystallization 74
number of contacts
   in a unit area of the failure section (χ) 198,
     202, 205, 211, 223, 225, 227, 237
   in the failure section 197

oil and gas fields 7, 77, 84, 122, 125, 131, 133,
   139, 188, 191
   Nizhnevartov 139–142
   Surgut 139, 141, 142
   Taz 139, 141
   Urengoy 139, 141
   Us't-Balyk 139
   Zapolyarnoe 139, 141
opoka-like clay 200
organic matter 3, 4, 11, 13, 31, 38–41, 50, 61,
   66, 67, 71, 72, 74, 85, 121, 123, 125–130,
   132–134, 137–139, 141, 147, 148, 152,
   164, 172, 182, 191
   admixture 128, 134, 137, 138, 147
   fine humus 130
   of terrestrial type 147
   oil- and gas-producing 125
   oil- and gas-prone 121
   phytogenous 123
   sapropel-humus 129, 134
   sapropel-algal 148
   terrigenous 137
organic remains 13, 40, 133
orientation rose of structural elements 34, 36,
   43, 45, 53, 65
origin
   lacustrine 13
   marine 13, 148
overpressure 47, 86, 117, 165

paleoclimatic zonation 123
paleodeltas of rivers 126
palygorskite 10
parameters
   initial 234, 235
   intervening 237
   micromorphological 226, 248–250
   microscopic 198, 211
   principal 74, 237
   structural-mechanical 197
particle size 30, 79, 92, 133, 134, 170, 214,
   215, 220, 226, 228

particles
   anisometric 32, 202
   clay 4–6, 9–11, 13, 22, 27–31, 33, 40, 42, 44,
     46, 51, 53, 55–60, 80, 85, 92, 94, 102, 116,
     118, 127, 131, 150, 152, 155, 164, 165, 170,
     171, 202, 204, 209, 211, 212, 214–218, 221,
     225–228, 235, 236, 248–250
   interacting 14, 27, 28, 56, 58, 206, 207
   mineral 3, 9, 24, 29, 41, 51, 61, 67, 90, 91,
     106, 151, 164, 206–208, 223
   spherical 18, 20, 24, 25, 30, 63, 198, 222
paste of clay 81
periclines 183
permeability
   clay 92, 94, 95
   diffusion-related 132
   filtration 117, 188
   filtration-related 132
   fracture-related 132
   of sediments 113, 188
   rock 91, 94, 95, 117, 118, 188
   seals 96
permeability index (K_p) 90, 93, 192
pH of the medium 5, 24
phase contact formation 59, 182, 207
phase of basin
   regressive 121
   transgressive 121
photomontage of SEM images 33, 43, 51, 53
phyllite-like slates 63
physicochemical mechanics 8, 109, 118, 197,
   229
physicochemical models of clay rock
       microstructure (see model) 197, 211
picoNewton 206
piezoelectric cell 206
piezoelectric displacement transducer 205
plane spring (see dynamometer) 205
plasticity 56, 182, 183, 185, 187, 191
platforms 46, 121, 122, 126, 183
   ancient 121
   epi-Hercynian 121
   young 121
point cementation 207
Poisson's ratio 58
polarizing action of counter-ions 87
polycrystalline aggregate 56
polycrystalline overgrowths 51
poorly dissoluble salts 3
poorly-oriented clays and clay rocks 17

pore channels 90–92, 94
pore solution 14, 41, 42, 46, 47, 61, 66, 69, 70, 85, 92, 96, 167, 180, 197, 212
pore space 13, 14, 34, 35, 42, 44, 51, 53, 54, 64, 75, 78–83, 86, 90–92, 94, 95, 114, 118, 166, 172, 189, 192, 212
pore space structure 79, 80, 81, 90, 95, 96, 114, 115, 117, 168, 188, 189
pores
　anisometric 14, 16, 35, 44, 51, 52, 55, 79, 80, 166
　biogenetic 14–16
　fissure-like 14, 16, 51, 53, 55, 64, 96
　interaggregate 14–16, 42
　intergranular 14–16
　intermicroaggregate 14–16, 33, 35, 42, 44, 51, 52, 54, 55, 79–81, 95, 96, 113–115, 117, 166, 168, 174
　intermicroaggregate-granular 14–16
　interparticle 14, 15, 35, 44, 45, 53, 55, 64, 79–81, 113, 117
　interultramicroaggregate 14–16, 35, 44, 45, 51, 52, 55, 79–81
　intragranular 14–16
　isometric 14, 16, 33, 35, 44, 51, 79, 80, 166
pores-cells 33, 113, 166–168, 177
porosity 8, 13, 30, 39, 42, 44, 58, 63, 64, 66, 67, 71, 72, 74–77, 79–84, 90, 94–97, 102, 114, 117, 132, 133, 164, 166–168, 170–174, 176–178, 180, 189, 191, 193, 198–202, 226, 227, 232, 233, 236, 249
　active 14, 91, 94, 96, 189
　clay 67, 69, 76, 86, 172, 176
　closed 13, 74
　effective 13, 14, 74, 189, 191, 192
　open 13, 74–78, 85, 197
　rock 65, 66, 67, 69, 70, 78, 79, 83, 95, 107, 115, 118, 168, 175, 178, 189, 227
　total 13, 35, 44, 51, 55, 68, 74, 75, 78, 96, 114, 173, 189, 191, 192, 240, 242, 244, 246, 251, 239
potential gap 57
potential minimum 23, 28, 111, 209
　close 23, 25, 26, 30, 31, 36, 164, 209
　distant 23–25, 28, 30, 32, 45, 164
precision 205–207
present day muds 203, 205
pressure
　anomalously high formation 47, 86
　anomalously high pore (over pressure) 117
　breakdown 95, 229
　capillary 75, 86
　confining 98
　geostatic 38, 46, 47, 55, 66, 67, 69, 70, 79, 81, 88, 91, 94–96, 118, 187, 221, 223
　gradient 14, 91, 94, 96
　gravitational 165, 168, 171
　hydrostatic 86, 94, 116, 189
　hydrodynamic 91, 117
　pore 42, 86, 105–107, 114, 117
principle of actuality 137
probability density 35, 44, 52, 53
process
　argillization 80, 96, 97, 118, 178, 191
　biochemical 3, 38, 40, 165
　chemical 125
　coagulation 4, 24, 26, 27, 32, 92
　geodynamic 183
　hypergene 123, 125
　microbiological 38
　physical 123
　physical-mechanical 3, 46
　physicochemical 3, 38, 46, 63, 79, 86, 112–114, 123, 165, 168, 172
　physiogeographic 127
　recrystallization 46, 70, 74, 80–82, 96, 97, 118, 148, 152, 156, 178, 181, 191
　rheological 98, 109, 168, 183, 192
　thermodynamic 66, 86, 88, 173, 183, 189
program control buttons 234
properties
　elastic-viscose 29, 56
　filtration 7, 85, 90, 190
　isolating 7, 117, 150, 166, 185, 186, 191–193, 231, 239
　mechanical 26, 51, 197, 211
　of clay rocks 8, 17, 51, 66, 71, 86, 87, 181
　physical 51, 64, 71, 100
　physical-mechanical 6, 7, 97
　rheological 109, 111
　screening 13, 79, 87, 96, 97, 100, 116–118, 125, 126, 131, 133–135, 137–139, 147, 148, 150–152, 154, 155, 158, 172, 180, 188–193, 229, 231, 232, 255, 257
　structural-mechanical 109, 163, 165
　thixotropic 45, 221
　viscous-plastic 94
protonization (dissociation) of water molecules 87
protonization of forming free water 117

pseudocrystals 169, 184
pyrite 12, 13, 38, 74
pyrophyllite 64

quantitative analysis of the pore space structure
    79
quartz 3, 13, 19, 21, 30, 47, 60, 205, 208,
    221–223, 227
quartz microsphere 206
quartz plate 205
quarz spherical lens 205
quiet tectonic epochs 125

radiolarians 3, 13
ratio
    of exchangeable Na^+/Ca^{2+} 133, 134, 165
    of sand-silt fraction (sa) to the clay fraction
        (c) ($K_{sa/c}$) 148, 149, 153, 155, 156, 159,
        242, 244, 246
    of swelling to non-swelling minerals (K_m)
        149, 152, 165, 240, 242, 244, 246
Rebinder's effect 87, 88, 112, 117, 169
recrystallization of minerals 46
relatively deep-water part of shelf 125
relief of territories 123
reorientation of particles 94
repulsion barrier 23
reservoirs 121, 122, 132, 171
retarded elasticity 102
reversibility of transitional contacts 56, 57
rheology 109
rigidity 172, 189, 207, 226
rock dehydration 47, 82, 84, 86, 88, 97, 116,
    123, 185
rock type 137, 240, 242, 244, 246
rocks
    acid 123, 124
    basic 123, 124
    igneous 123
    metamorphic 123, 125
    sedimentary 37, 38, 45, 46, 100, 123, 125
    ultrabasic 123, 124

sample recovery 211
sampling 239
sand 3, 11, 16, 38, 41, 43, 58–60, 74, 80, 85,
    112, 121–123, 125, 127–135, 138–145,
    147–149, 152, 153, 155, 158, 159, 163,
    166, 168–171, 179, 182, 183, 185, 189,
    192, 200

sandstone 128, 132, 140, 141, 143, 183, 200
sapropel organic matter of algal type 138
sapropel 40, 67, 128, 133, 134, 152
schistosity 64, 70, 81, 118, 178, 184
Seal Test program 233–236, 249, 251–257
seals
    clay (see clay seals)
    deep-water shelf 142
    local 132
    middle-depth shelf 142
    of classes I–II 135, 138, 139, 142, 143, 145,
        148–151, 158, 159, 161, 203, 212–217,
        219–221, 225–235, 239, 248, 249, 251,
        253, 255, 257
    of classes III–IV 135, 138–143, 145, 146,
        152–154, 158–161, 186, 203, 212–216,
        219–221, 225–233, 235, 239, 248, 250–257
    of classes V–VII 135, 138, 139, 141–143,
        155–161, 169, 204, 219, 221–223,
        226–233, 235, 236, 239, 248, 249,
        251–254, 256, 257
    regional 122, 132
    sandy-silty clay 132
    semi-seals 132
    shallow-water shelf 142
    true 132
    zonal 132
section
    sedimentary 48, 121, 140
    sedimentary basin 122
    stratigraphic 139
sediments
    clay (see clay sediments)
    deep-water 42, 141, 147, 155, 165, 166, 168,
        177, 178, 180, 188, 192
    gas-prone 130
    Jurassic 139, 143
    river-bed alluvial 122
    unconsolidated 37
sediment accumulation 38–40, 46, 125, 131,
    132, 137–141, 145, 172, 183, 211
sediment microstructure (see microstructure)
    large-honeycomb 165
    medium-honeycomb 165
    small-honeycomb 165
sedimentary cycles 121–123, 130
sedimentary formations 137
sedimentary rock 37, 38, 45, 46, 100, 123, 125
sedimentogenesis 3, 23, 30, 74, 79, 119, 123,
    131, 137, 170, 211

SEM images 33, 43, 44, 51, 53, 54, 79, 226, 249
SEM photographs 208, 249
sequences
   carbonate 122
   sand-silt 122
   terrigenous 122
sericite 64, 68, 74
shales 63, 69, 130, 132
shallow-water coastal and shelf areas 125
shear rate 110, 111
shearing
   consolidated drained 106, 107
   consolidated undrained 106, 107
   plane (direct shear test) 104
   resistance 104
   under triaxial compression (triaxial test) 104
shelf (see environment)
   deep water 85, 125, 126, 128, 133, 139, 141,
      143, 147, 148, 159, 163, 179–181, 185,
      188, 193
   middle 128, 133, 135, 140, 144, 151, 166,
      168, 176–178, 181, 185, 186, 188, 190, 193
   shallow 128, 134, 135, 169, 170, 179, 181,
      187, 193
siderite 3
silts 130, 134, 138, 143, 155, 169, 171, 221,
   222, 225–228, 236, 242, 244, 251
siltstones 13, 63, 70, 98, 100, 103–105, 118, 128,
   140, 141, 143, 200, 204, 221–223, 226, 231
simple cubic packing 198
slates 63
Slichter's number 90
smectites (see montmorillonite) 9, 11, 38
sodium chlorite 207
solid structural elements 9, 13, 16, 33, 43, 51,
   53, 96, 97, 197, 216
salinity of filtering fluid 92
solution of calcium sulfite 207
source area 123, 127, 130, 131, 137, 138, 140,
   148, 150, 156
specimen chamber of SEM 207
spontaneous condensing of sediment structure
   113
stabilization of clay particles 42, 164
stage
   of combined development of geostatic and
      recrystallization compaction 82
   of free geostatic compaction 81, 82 178
   of impeded geostatic compaction 82, 175, 178
   of recrystallization compaction 82, 83

STIMAN 212, 249
strain
   elastic 76, 100, 115
   irreversible residual 102, 103
   relative elastic (ε_e) 100, 101
   relative volume (ε_v) 98
   total (ε_t) 100, 101
strength of individual contacts (P_1) 30, 204,
   205, 211, 212, 216, 219–223, 225
strength
   compressive (P_c) 229, 231, 233, 234, 236,
      239, 253
   maximum (peak) (τ_{max}) 107, 109
   of fine grained structure (P_s) 197
   residual (τ_{min}) 108, 109
   sample 102
   shearing 104, 107, 114, 165
   structural 70, 99, 100, 102, 105, 106, 169, 179
   tensile (P_t) 30, 58, 62, 63, 205, 211,
      229–231, 234, 236, 248, 249, 251, 252
   ultimate shearing 6, 108, 111, 112
   uniaxial compressive 100–103, 109, 114–116,
      118, 239, 240, 242, 244, 246, 248
strengthening of structures bonds 46
stress 45, 47, 56, 61, 63, 64, 80, 96, 101, 103,
   104, 108–114, 182, 183
   critical 59
   effective 45, 58, 86, 106, 109, 114, 165, 171,
      180, 221
   effective external 58
   extensive 183
   external 58, 111, 182, 183
   mechanical 112
   natural 109
   normal 104, 105
   shear (tangential) 104
   shearing 82, 106, 108–113
   tectonic 183
   total 114
stress concentration 82, 86, 87, 96, 112, 117,
   165, 172, 178, 180, 182
stress dissipation 183
stressed state 71, 98, 182, 183
structural bonds 8, 9, 26, 28, 29, 46, 56, 64, 66,
   67, 69, 80, 81, 87, 88, 97, 98, 103, 104,
   106, 107, 111, 114, 115, 168, 169, 175,
   178, 180, 182, 184, 207
structural coherence 29, 113, 178
structural component of the wedging pressure
   18, 19, 21, 22

structural elements 8, 9, 11, 14, 16, 17, 23, 26, 27, 35, 46, 47, 52–56, 63, 64, 69, 70, 79, 80, 82, 94, 95, 97, 102, 105, 106, 108–114, 178, 179, 182, 183, 185, 191, 199–201, 203, 204, 212, 225, 226, 231

structural parameter (N) 198, 232

structural rearrangement 87, 106, 118
  of particle 106, 107, 114
  of rock 117

structure formation 3–5, 7, 8, 17, 23, 30, 32, 163, 166, 169
  thixotropic 111

structure 4, 5, 7–10, 12, 16, 19, 24–27, 32, 33, 40–42, 44, 51, 53, 56, 57, 60, 61, 66, 67, 70, 71, 79, 80, 81, 83–86, 89, 91, 95, 98, 108–113, 115, 116, 123, 126, 131, 133, 134, 139, 144, 163, 165–169, 172, 173, 175–177, 180, 182, 183, 191, 192, 198, 199, 203, 205, 209, 211, 221, 231
  atomic 7, 9
  conformal-regeneration 180
  clay 5, 7, 8, 12, 13, 16, 60, 66, 86, 96, 131, 135, 138, 203, 204, 226
  crystal 7
  crystallization 8, 221
  crystallization-cementation 231
  crystallochemical 90, 92
  cyclic 121
  fine grained 26, 131, 197, 198, 201
  of a sediment sequence (facies, formations) 7
  of the Earth crust 7
  of the Earth 7
  rhythmic 122
  rock 7–9, 80, 97, 117, 118, 183, 188
  swelling crystalline 121

subaerial delta plain 127

subaqueous lithogenesis 38, 72, 82, 83, 99

sulfides 38, 74

summary energy curve 22, 23

summary forces of partial interaction 21

surface
  basal 5, 25, 28, 32, 35, 53, 56, 59, 209, 210, 216
  crossed cylindrical 207, 209
  hydrophobic glass 206
  interphase 197
  mineral 5, 6, 19, 28, 40, 47, 94, 167, 182
  silicate 6
  specific 90, 92, 97
  unit 4, 14, 27, 151

surface tension 75

swelling 33, 40, 48, 50, 56, 59, 74, 84, 85, 87, 121, 125, 134, 135, 147, 149, 150, 152–154, 157, 159, 160, 163–166, 168, 169, 173, 185, 189, 212, 214–216, 240, 242, 244, 246

swelling crystal lattice 135

swelling layered silicates 125

syneresis 39, 41, 113

system
  bi-dispersed 201
  dispersed 24, 167
  fine grained 8, 17, 21–25, 27, 30, 63, 113, 197, 198, 225
  heterogeneous multicomponent 197
  magnetoelectric 206, 207
  polydispersion 24, 171
  river 127

talc 207, 223, 224

tectonic fracturing 97

temperature 5–7, 18, 24, 38, 45–47, 55, 56, 59, 60, 64, 66–71, 80, 83, 85–89, 91, 94–98, 100, 105, 109, 112, 116, 117, 118, 148, 165, 172, 182, 183, 185, 188, 191, 197, 207, 225

tensile strength 30, 58, 62, 63, 205, 211, 229–231, 234, 236, 248, 249, 251, 252

tensile test 205

testing
  strength 101, 103, 105
  under the loading–unloading regime 101, 102

tests
  compression 98, 99, 101, 107
  ring shear 107
  shearing 104, 107

texture 8, 9

theory of mineral dispersion 4, 5

thermodynamic dehydration 191

thermodynamic equilibrium 41

thermodynamic nature 31, 86, 113, 116

thermodynamic removal 189–191

thixotropic restoration 109, 113

thixotropy 41

tides 127

transfer
  diffusional 85
  filtration 86, 117, 165, 191
  mass 7, 86, 88, 116, 168
  moisture 165
  osmotic mass 86, 88, 116

transformation
  lithogenetic 38, 48, 69, 148, 212
  of structure 46
  pore space 79, 81, 83
  sediment 114, 133
transgression 121, 122, 126–128, 130, 139,
    141–146
  maximum 126, 127, 130, 141, 145, 146
  middle 126–128, 142
transgressive epochs 122
transportation 40, 123
triaxial compression 98, 100, 104
turf 40, 67

ultramicroaggregates 9, 10, 11, 12, 14, 16, 24,
    32, 33, 36, 44, 51, 53, 55, 164, 168
ultramicropores 14, 35, 44, 45, 51–53, 55, 64,
    75, 79, 80, 85, 86, 113, 116, 117, 164, 188,
    189, 191
uncompensated valence bonds 4
uniaxial compression 97

varved clay 205
vegetable humus 139
vermiculite 125, 207, 223, 224
viscosity 14, 47, 83, 86, 87, 90, 94, 95, 102,
    110–116, 168, 171, 175, 191
  differential 110
  plastic 111, 112
  structural 171
void ratio 75, 98, 199

water
  adsorptionally bound 5, 6, 42, 47, 56, 69, 84,
    86–88, 94, 96, 116, 117, 165, 173, 175,
    178, 183–185, 188–192
  capillary 5, 7, 14
  chemically combined 7
  distilled 93, 216
  free 5, 6, 19, 42, 46, 47, 66, 67, 69, 82,
    84–88, 92, 94–96, 102, 113, 115, 117,
    168–171, 173–175, 178, 183–185, 188
  loosely bound 7, 86, 88, 91, 94, 116
  ooze 40
  osmotic bound 6
  physically bound 5, 66, 84, 85, 86, 113
  squeezed-out 46, 47, 70, 85
  tightly bound 5, 47, 87, 88, 91, 93, 178

water column weight 42
water content 36, 40, 41, 67, 69–73, 83–86, 88,
    89, 91, 105, 106, 113–116, 166, 167, 170,
    197
  maximal hygroscopic (W_{mh}) 84
  natural 36, 89
  residual 88
  volume (W_V) 84
  weight (W) 71, 84, 116, 170
water molecules 5, 6, 19, 28, 42, 47, 85–88,
    116, 117
weathering
  ancient crusts 123, 124
  crusts 3, 123, 125
  crust horizons 125–127, 129, 150
weathering crust profile 123, 125, 130
weathering products 123
weight content 84, 227, 228, 249
West Siberia 48, 49, 76, 78, 121, 122, 126, 131,
    132, 139, 141–143, 173, 174, 176
West-Siberian lowland 122, 132

zeolites 38
zone
  arid (see arid zone)
  catagenesis 45–47, 70, 79, 100, 107, 112
  climatic 123
  coastal 127, 130, 131, 142, 147, 155, 159,
    169, 189
  deep water 76, 147, 150, 183
  desintegration 124
  diagenesis 73, 74, 79, 98, 171, 178, 182, 192
  humid 48, 49
  hydrodynamically quiet 113
  hypergenesis (see weathering) 123, 125, 126
  leaching 124
  lithogenesis 66, 72
  littoral 42, 76
  metagenesis 118
  microfracture 183
  near-delta 131
  of anomalous horizons 96
  shallow-water 41, 142, 179, 181, 187
  shallow-water coastal 127, 143
  shallow-water shelf 42, 76, 142
  shear 102, 106–108
  shelf sedimentation 127
  weathering 123, 127